Essays on Supersymmetry

MATHEMATICAL PHYSICS STUDIES

A SUPPLEMENTARY SERIES TO
LETTERS IN MATHEMATICAL PHYSICS

VOLUME 8

Essays on Supersymmetry

Edited by

C. FRONSDAL
Department of Physics,
UCLA, Los Angeles, U.S.A.

With contributions by

M. FLATO
Professor of Mathematical Physics,
Dijon University, Dijon, France

C. FRONSDAL
Professor of Physics,
University of California, Los Angeles, CA, U.S.A.

T. HIRAI
Professor of Mathematics,
University of Kyoto, Kyoto, Japan

D. Reidel Publishing Company

A MEMBER OF THE KLUWER ACADEMIC PUBLISHERS GROUP

Dordrecht / Boston / Lancaster / Tokyo

Library of Congress Cataloging in Publication Data

CIP

Essays in supersymmetry.

(Mathematical physics studies; v. 8)
Bibliography: p.
Includes index.
1. Supersymmetry. I. Fronsdal, Christian, 1931- II. Flato, M. (Moshé), 1937- . III. Hirai, T. (Takeshi), 1936- IV. Series
QC174.17.S9E87 1986 530.1 86-602
ISBN-13: 978-94-010-8555-7 e-ISBN-13: 978-94-009-4624-8
DOI: 10.1007/978-94-009-4624-8

Published by D. Reidel Publishing Company,
P.O. Box 17, 3300 AA Dordrecht, Holland

Sold and distributed in the U.S.A. and Canada
by Kluwer Academic Publishers,
190 Old Derby Street, Hingham, MA 02043, U.S.A.

In all other countries, sold and distributed
by Kluwer Academic Publishers Group,
P.O. Box 322, 3300 AH Dordrecht, Holland

Softcover reprint of the hardcover 1st edition 1986

This book is dedicated to Amelita.

CONTENTS

INTRODUCTION

1. Why supersymmetry? 1
2. Why in de Sitter space? 3
3. Why group theory? 5
4. Background. 5
5. This book, summary. 8
6. Future directions. 11

UNITARY REPRESENTATIONS OF SUPERGROUPS
C. Fronsdal and T. Hirai

0. Introduction. 16
1. General structural problems. 19
2. Invariant Hermitean forms. 33
3. An example: osp(2n/1). 53

3+2 DE SITTER SUPERFIELDS
C. Fronsdal

0. Introduction. 68
1. Superfields and induced representations. 69
2. Induction from an irreducible representation. 71
3. Invariant operators. 79
4. Massive superfields, "scalar" multiplet. 84
5. The "vector" multiplet. 88
6. The simplest superfield for N = 2 supersymmetry. 94
7. Induction from an irreducible representation. 98
8. Wave equations for N = 2. 104
9. The spinor superfield and de Sitter chirality. 106

APPENDICES

A1. Linear action for osp(2n/1). 111

A2. Linear action for osp(2n/2). 114

A3. Intertwining operators. 116

A4. Invariant fields. 117

SPONTANEOUSLY GENERATED FIELD THEORIES, ZERO-CENTER MODULES, COLORED SINGLETONS AND THE VIRTUES OF N = 6 SUPERGRAVITY

M. Flato and C. Fronsdal

0. Introduction. 124
1. De Sitter electrodynamics. 136
2. Conformal electrodynamics. 139
3. De Sitter super electrodynamics. 140
4. Extended de Sitter super electrodynamics. 144
5. Super conformal electrodynamics. 151
6. Extended super conformal electrodynamics. 154

MASSLESS PARTICLES, ORTHOSYMPLECTIC SYMMETRY AND ANOTHER TYPE OF KALUZA-KLEIN THEORY

C. Fronsdal

0. Introduction. 164

I. Geometric preliminaries.

1. Phase space, sp(2n,R) and osp(2n/1). 167
2. The oscillator representation. 170

II. Superfield preliminaries.

3. Superfield representation. 174
4. Oscillator representation on superfields. 177

III. Algebraic representation theory.
5. Lowest weight representations of sp(2n). 179
6. K-structure. 185
7. Non-decomposable representations of sp(2n). 188
8. Lowest weight representations of osp(2n/1). 191
IV. Homogeneous space and line bundle.
9. The homogeneous space X. 195
10. Parameterization of X. 199
11. The line bundle Z^{α} over X. 202
12. The oscillator representation on $Z^{-1/2}$. 206
V. Physical interpretation.
13. The conformal group $U^{\gamma}(n)$. 210
14. Identification of space time. 215
15. The other orbits of $U^{\gamma}(n)$. 218
16. Interpretation of the extra dimensions. 222
17. Irreducible $u^{\gamma}(n)$ modules. 227
VI. Scalar field on space time.
18. Scalar field on Dirac's projective cone. 231
19. Quasi-invariant wave operator on U(2). 236
20. The conformal tube. 239
21. Hilbert space of holomorphic functions. 243
22. Invariant bilinear functionals. 247
23. Cauchy kernel and Lagrangian. 252
VII. osp(8) field theory--a beginning.
24. Intertwining operators. 256
25. Lagrangian and wave equation. 261
26. The reduced superfield. 262

INDEX 266

"one has no clue to what new physical ideas are needed. However, one can be sure that the new theory must incorporate some very pretty mathematics, and by seeking this mathematics one can have some hope of solving the problem."

P. A. M. Dirac, *Proc. Roy. Soc.* **A223**, 438 (1954).

"I once spent part of an afternoon trying to explain $p^2 : 2pq : q^2$ to William Bateson. Not only could he not understand it but he could see no possible point in it." E. B. Ford, quoted by R. W. Clark in *The survival of Charles Darwin*, Random House, N.Y., 1984.

INTRODUCTION

1. Why supersymmetry?

Supersymmetry has yet to find its proper place in physics—so why do we believe in it?

Our concept of space time was shaped by special relativity, only to be thoroughly confused by the general theory. First there was the problem of large distances associated with the cosmological constant.[1] Infrared problems of electrodynamics also relate to large distances, while ultraviolet divergences raise questions about the structure of space time at very small distances. It must be said that the boundaries of space time at finite distances also are imperfectly understood, as in the case of the Schwartzschild solution.[2] General relativity also raises questions about the observational meaning of space time events. All of these problems have contributed to a feeling of dissatisfaction, as can be seen from the numerous attempts to find a deeper level of understanding, in terms of new structures underlying our concepts of space and time. It is an indisputable merit of supersymmetry that, of all proposals that have been made so far, to alter or to develop our views on space time structure, it comes closest to proving its physical relevance.

The hope of unifying the fundamental forces of nature is a central theme of particle physics. The degree to which unification is achieved in the electroweak theory of Weinberg[3] and Salam[4] can be debated—what is the contribution of supersymmetry?

Forces are usually thought of as the exchange of bosons; it is not obvious, therefore, that one needs a symmetry that relates bosons to fermions. But supersymmetry offers the hope of uniting all particles, perhaps in a single irreducible representation. Of course, this can happen only in some ideal or primordial world; the relation of that world

C. Fronsdal (ed.), Essays on Supersymmetry, 1–14.

to our own also needs to be understood. Such unification may also require that the supersymmetry group possess irreducible representations with infinite reduction on the Poincaré subgroup, to accommodate an infinite set of particles. Such possibilities were envisaged long ago[5] and have recently reappeared in Kaluza-Klein supergravity[6] and in superstring theory.[7] Supersymmetry implies that forces that are mediated by bose exchange must be complemented by forces that are due to the exchange of fermions. The masslessness of neutrinos is suggestive—we continue to favor the idea that neutrinos are fundamental to weak interactions, that they will finally play a more central role than the bit part assigned to them in Weinberg-Salam theory.

There seems to be little room for doubting that supersymmetry is badly broken—so where should one be looking for the first tangible manifestations of it?

It is remarkable that the successes that can be legitimately claimed for supersymmetry are all in the domain of massless particles and fields. Supergravity is not renormalizable, but it is an improvement (in this respect) over ordinary quantum gravity. Finite super Yang-Mills theories are not yet established, but there is now a strong concensus that they soon will be. In both cases massless fields are involved in an essential way. It seems that supersymmetry, masslessness and conformal invariance are the fundamental principles from which finite field theories must be constructed, although all three have to be partly or wholly abandoned in the final and physically most relevant formulation. The success of electroweak gauge theory is yet to be generalized, but it surely is a valuable paradigm. An even more fundamental point is that massless field theory is the common ground of all radical attempts to modify space time structure. We have in mind Penrose's twistor program, as well as Kaluza-Klein theories and (super) string models.

2. Why in de Sitter space?

One of the unusual features of the papers presented here is the emphasis on de Sitter field theories. The usual reaction to the cosmological term is to regard it as a nuisance—but this is most decidedly a mistake.

A fundamental reason for taking the cosmological constant (henceforth: Λ) into consideration is that there is no good reason to expect it to be zero, and no hope of ever finding it to vanish by observation. Equally important is the fact that, within the family of spaces with maximal symmetry, the flat space value $\Lambda = 0$ is highly singular. This means that interesting and important structures may exist when $\Lambda \neq 0$ that have no meaning in the limit $\Lambda = 0$. This is indeed the case, especially in the context of gauge field theories. The most spectacular instance is the existence of two very remarkable elementary objects (the Di and the Rac, collectively singletons),[8] that can be interpreted as the constituents of massless particles, and perhaps identified with quarks. Among the exceptional properties of singletons we recall that two-singleton states are always massless (a purely kinematical fact), and the circumstance that they are described by gauge fields (though the spins are 0 and 1/2).[9] It is necessary to explain precisely what we mean by de Sitter space and (super) de Sitter symmetry.

It is crucial to choose the good de Sitter group and the good de Sitter space. The 4+1 de Sitter group was favored at first, but it can be excluded because it has no unitary representations with positive energy, and hence no particles. This leaves the 3+2 de Sitter group (henceforth: <u>the</u> de Sitter group). The Lie algebra is isomorphic to sp(4,R), and the graded extension is osp(4/1). Now 3+2 de Sitter space can be identified with (a covering of) the hyperboloid $y_0^2 + y_5^2 - \vec{y}^2 = 1/\rho$ in R^5. The constant ρ must be positive, for only then does the space admit an invariant causal structure. This was shown by Castell in 1969.[10] Since this is the only one of the de Sitter spaces that is physically interesting, we have always referred to it simply as de Sitter space. [The term anti-de Sitter space was introduced later by Ellis and Hawking,[11] for the unphysical version with negative ρ.] We can now return to the comparison between physics in de Sitter space and physics in flat space.

Ultraviolet renormalization has long been regarded as the main problem with relativistic field theories, but recently it has become clear that the infrared catastrophe is also worthy of attention, perhaps even more so. The best available treatment of the infrared regime in ordinary QED is not above criticism.[12] The foundation of QCD is the hope that catastrophic infrared behavior may be interpretable as confinement. Dynamical breakdown of supersymmetry is also strongly related to infrared behavior.[13] One of the great merits of de Sitter field theories is the very effective infrared regularization that is introduced by the curvature. In some ways de Sitter space is like a box of volume $\rho^{-3/2}$. Actually, the density of levels near zero energy is much more rarified than in a box of finite volume, and yet this is brought about without any loss of symmetry. The scarcity of states is not the only feature that is characteristic of de Sitter field theories. There is also an interesting fine structure that is lost in the flat space limit. Many types of positive energy representations become massless in the limit, but a much smaller subset is related to gauge theories.[14] One understands, much more clearly than in flat space, that the essence of gauge theories is group theoretical: a relativistic field theory is a gauge theory if and only if the field module is nondecomposable. Yet another intriguing feature of massless theories in de Sitter space is the peculiar nature of helicity.[15] Strictly, helicity is not a group theoretical notion in de Sitter space. Consequently, weak interaction phenomenology is a little different from the familiar, flat space version, and not yet completely understood.

3. Why group theory?

The regularizing effect of supersymmetry depends on the imposition of exact invariance on a primordial, purely massless field theory. The appropriateness of the group theoretical idiom is self-evident—so why is it that group theory is seldom fully exploited in the formulation of massless quantum field theories?

The usual answer is that gauge invariance makes the application of group theory difficult. The physical, massless states of gauge field theories carry unitary representations of the symmetry (super) group, just as in the massive case, but the modes of free gauge fields carry non-decomposable representations with indefinite metric.[16] This complicates matters to some extent, but it certainly does not lessen the value of group theory, quite the contrary is true. One of the central problems of supersymmetric (or conformal) field theories is the discovery of the auxiliary fields that are required for quantization. An initial field structure is given, on which the symmetry (super) group acts by a nondecomposable representation D, say. Typically, D does not admit a non-degenerate symplectic form, and in that case invariant quantization is impossible. The problem is group theoretical: to find an extension of D that possesses a non-degenerate (indefinite) metric.

4. Background.

Elements of supersymmetry can be traced back to the work of Schwinger in 1953[17] on the quantum action principle, where anticommuting numbers were introduced for the first time. Invariance under a graded extension of the Poincaré group was first discussed by Flato and Hillion in 1970.[5] The superalgebra that has become most closely associated with the term supersymmetry was discovered by Gol'fand and Lichtman in 1971.[18] They constructed the basic representations (the scalar and vector multiplets) of this algebra by induction from the even part, and investigated invariant Lagrangians. Wess and Zumino, in 1973,[19] improved on this work and did much to call attention to the exiting prospects of these supersymmetric field theories. Salam and Strathdee[20] introduced the all-important method of superspace and superfields. One of the most significant advances in supersymmetry, the supersymmetric extension of QED and Yang-Mills theories, was achieved by Wess and Zumino, Ferrara and Zumino, and by Salam and Strathdee, in 1974.[21]

The possibility of generalizing all aspects of modern particle physics to de Sitter space has occurred to many people, as much as 50 years ago.[22] Our own participation in this program goes back to 1964.[23] We advanced arguments that point to 3+2 as the most promising signature, pointed out the particle interpretation of positive energy representations, and began a rough classification of the unitary, irreducible ones (in terms of the highest weights). Field quantization was taken up in 1974,[24] based on the causal structure discovered by Castell.[10] Next, we call attention to the results of a paper[14] published in 1975. That paper studied fields with spins 0, 1/2, and 1, especially the massless case. The most significant results are as follows:

(a) For each spin the field equations admit two sets of solutions, with each set carrying a unitary, irreducible representation of SO(3,2).

(b) For spin zero, the two representations are inequivalent, and the wave functions are not mutually orthogonal. This was interpreted in terms of the Hamiltonian. If both sets of modes are carried by one and the same field, then the Hamiltonian is not self-adjoint.

(c) For spin 1/2, basically the same result is obtained, though in this case the two representations are equivalent. The "massless" case is identified by conformal invariance. In this particular case, there exists an invariant matrix function that generalizes the helicity matrix γ_5 to de Sitter space in a natural way.

(d) For spin 1, the massless case is identified in several different ways, notably by conformal invariance and the breakdown of the Fierz-Pauli program. The latter is tantamount to the appearance of nondecomposable representations, and gauge theory.[14] [Analogous results for spin 2 were obtained later,[25] and the general case of integer or half-integral spins was also treated, though in less detail.[26]] We observe that the relationship between mass and cosmological constant, the meaning of masslessness, and the correct definition of chirality, were all pretty well understood at that time, though not very widely known even at present.

The extension of supersymmetry to de Sitter space, or of de Sitter field theory to include supersymmetry, was first considered by Keck in 1975.[27] Like Gol'fand and Lichtman,[18] and Salam and Strathdee,[20] he constructed representations of the graded algebra osp(4/1) by induction from the even part. A very simple formulation of de Sitter supersymmetry was given soon afterwards.[28] The notation for de Sitter and for conformal supersymmetry can actually be made more concise than Poincaré supersymmetry, although one often gets the contrary impression.

The importance of an efficient notation was already recognized by Leibniz,[29] to whom we owe some of the conventions of modern mathematics. As a rule, notation develops as needed to express new ideas, but supersymmetry is an exception. [We admit that supersymmetry contains new ideas.] Much writing on supersymmetry employs a notation that makes us think of the Alexandrian school of arithmetic.[30] As an example of a more efficient notation we offer the following. Let $\{\Omega_{\dot{a}b}\}$ a,b = 1,...,4 be the basis for an operator representation of u(2,2), with commutation relations

$$[\Omega_{\dot{a}b},\Omega_{\dot{c}d}] = \delta_{b\dot{c}}\Omega_{\dot{a}d} - \delta_{d\dot{a}}\Omega_{\dot{c}b} \ .$$

Here δ denotes the metric that defines u(2,2) in $g\ell(4)$. Let $\{\theta^a\}$ a = 1,...,4 be the generators of a Grassmann algebra. Then a realization of the graded algebra su(2,2/1), by operators acting on superfields, is given by

$$q_a = \partial_a \equiv \partial/\partial\theta^a \ , \quad q_{\dot{a}} = \theta^b(\Omega_{\dot{a}b} - \delta_{\dot{a}b}\theta\cdot\partial) \ ,$$

$$T_{\dot{a}b} = \theta_{\dot{a}}\partial_b + \Omega_{\dot{a}b} - \delta_{\dot{a}b}\theta\cdot\partial \ .$$

These simple formulas should be compared with the complicated and particularized expressions that are found in most papers on conformal supersymmetry. Note that one rarely needs to know the explicit form of $\Omega_{\dot{a}b}$ as vector fields on space time.

5. This book, summary.

"Unitary representations of supergroups" is an attempt to begin the systematic study of unitarizable representations of noncompact superalgebras. There is at present no clear concensus on how to define supergroups. The approach taken here is that of physicists; we do not attempt to convert supergroups to groups, because we feel that the interesting structure of superalgebras is lost by transferring the emphasis to the affiliated Lie groups. We first define the local supergroup associated with any Lie superalgebra, with the aid of an auxiliary Grassmann algebra. These local supergroups are next shown to have the structure of semi-direct products, $G = G_o \cdot X$, in which G_o is a local Lie group and X consists of polynomials in the universal enveloping algebra of the original Lie superalgebra. A supergroup is defined to be a semidirect product $G = G_o \cdot X$, where G_o is now a Lie group. Definitions of representations, integrability and unitarizability follow quite naturally. Representations of superalgebras, induced from representations of the even part, are studied systematically and lead to a natural definition of "functions on supergroups:" they are functions on the Lie subgroup, with values in the vector space dual of the exterior algebra over the odd part of the superalgebra; that is, they are superfields. The principal new result is a proof that an invariant (Berezin) measure exists for integrating such functions. This measure includes the Haar measure on the Lie subgroup as a factor. With the aid of the invariant measure we prove that if π is a representation of a superalgebra, induced from a representation π_{in} of the even part, and if π_{in} admits an invariant, nondegenerate hermitean form, then so does π. This is the crucial property that guarantees that all necessary auxiliary fields are included in the superfield. [It is not true that every unconstrained superfield realization automatically satisfies this condition.] In the special case of osp(2n/1) we calculate the invariant Berezin measure explicitly.

"3+2 de Sitter superfields" examines the reduction of de Sitter superfields into irreducible representations of osp(4/1). New results include a detailed study of the group theoretical content in the massless case. By this we demonstrate (a) that the group theoretical analysis is even more useful in the massless case than in the more straightforward massive case and (b) that orthosymplectic symmetry is easier to handle than the singular super Poincaré limit. The various sub- and quotient spaces that enter into the analysis of indefinite metric, Gupta-Bleuler quantization are defined in terms of the "Dirac operator," a natural generalization of the operator $\overline{D}D$ of Poincaré supersymmetry. The geometrical structure of de Sitter supersymmetry includes a 5-dimensional derivative and an invariant 5 by 5 matrix superfield that plays the role of γ_5 in the definition of de Sitter chirality. We emphasize that de Sitter chirality is an exact notion; chiral and antichiral fields do exist, other claims notwithstanding. The simplest aspects of extended N = 2 de Sitter supersymmetry are also examined. A comparison with ordinary conformally invariant field theories gives an interesting view of the difficulties involved in the quantization of N = 2 supersymmetric field theories.

"Spontaneously generated field theories, zero-center modules, colored singletons and the virtues of N = 6 supergravity" introduces the vacuum ghost. Massless field theories that are invariant under either the conformal group or the de Sitter group, or the associated supergroups, contain a new type of ghost in the physical sector. The most favorable situation is realized in the several extensions of electrodynamics, for in these theories the ghost has only one degree of freedom, with the quantum numbers of the vacuum. We argue that this vacuum ghost is beneficial, that it helps to regularize the theory in the infrared domain, and that it may be associated with the Higgs-Kibble mechanism. There is only one theory in this favored class that includes gravitation (without higher spins), and that is N = 6 supergravity.

"Massless particles, orthosymplectic symmetry and another type of Kaluza-Klein theory" was originally inspired, in part, by the twistor program, and by our attempts to construct massless field theories with unbounded spins. However, we here summarize the results from the point of view of Kaluza-Klein. There exists a real 10-dimensional space, on which the group Sp(8,R) acts homogeneously. The singleton (oscillator) representation of osp(8/1) is realized in a natural way on superfields defined on this space. Restricted to the even part, it integrates to the sum of the two singleton representations of Sp(8,R). Reduction on the subgroup SU(2,2) (the ordinary conformal group) gives the direct sum of all the massless representations, each with multiplicity 1. Unlike other Kaluza-Klein theories there are no massive states. The action of the conformal group in the 10-dimensional manifold leaves invariant a unique 4-dimensional submanifold, that can be identified with (conformally compactified) space time. Dimensional reduction thus yields a conformal field theory on Minkowski space, in which the higher Kaluza-Klein excitations are all massless, but with unbounded spins.

6. Future directions.

To justify future hopes one has to acknowledge past failures. In the domain of naive phenomenology (particle classification and spectra), there is not a shred of evidence for supersymmetry. This was in fact obvious from the beginning, and recent last-ditch attempts to discover supersymmetric partners in weak boson production experiments can only bring discredit to supersymmetry. The successes of supersymmetry are formal and (one would like to say) of a fundamental nature. Supersymmetry reminds us of conformal invariance: it has profound implications for the structure of space time at very small (and perhaps very large) distances, for scaling phenomena and for the coherence of field theories, but its relevance for the understanding of particle spectra is still an open question.

We believe that the best prospects for supersymmetry lie in the further development of fundamental ideas. This does not necessarily imply that the goals have to be shrouded in the mists of a distant future. On the contrary, there are exciting ideas that may soon lead to very substantial progress. The most basic of all supersymmetry multiplets is the Dirac super-singleton,[31] consisting of one "boson" (the rac) and one "fermion" (the di). These truly remarkable objects are the constituents of massless particles. (All two-particle states are massless!)[8] One particle states are unobservable ("confined") for reasons that are purely kinematical, and three-particle states are massive. With the incorporation of unconventional statistics these objects become the quarks of strong interaction physics.

ACKNOWLEDGEMENTS

I wish to express my deep gratitude to Gilda Reyes for the dedication and perseverance that she has applied to the preparation of this book. I also thank George Hockney for his friendly help in the use of his versatile **Egg** word processing program, Lisa Capuano for her patient editorial work and the National Science Foundation for financial support.

References.

1. A. Einstein, Preuss. Acad. Wiss., 1919, pp. 349, 463.
2. K. Schwartzschild, Preuss. Acad. Wiss., 1916, p. 189.
3. S. Weinberg, Phys. Rev. Lett. 27, 1264 (1967).
4. A. Salam, Proceedings of the 8th Nobel Symposium, Gotenburg, 1968.
5. M. Flato and P. Hillion, Phys. Rev. D1, 1670 (1970).
6. E. Cremmer and B. Julia, Nucl. Phys. B159, 141 (1979).
7. J. H. Schwartz, Nucl. Phys. B226, 269 (1983).
8. M. Flato and C. Fronsdal, Lett. Math. Phys. 2, 421 (1978); Phys. Lett. 97B, 236 (1980).
9. M. Flato and C. Fronsdal, J. Math. Phys. 22, 1100 (1981).
10. L. Castell, Nuovo Cim. A61, 585 (1969).
11. G. F. R. Ellis and S. Hawking, Large Scale Structure of Space Time, Cambridge University Press, 1973.
12. C. Itzyksohn and J.-B. Zuber, Quantum Field Theory, McGraw-Hill, N.Y., 1980.
13. E. Witten, Nucl. Phys. B202, 253 (1982).
14. C. Fronsdal, Phys. Rev. D12, 3819 (1975).
15. E. Angelopoulos, M. Flato, C. Fronsdal, and D. Sternheimer, Phys. Rev. D23, 1278 (1981).
16. F. Strocchi and A. S. Wightman, J. Math. Phys. 15, 2198 (1974). G. Rideau, J. Math. Phys. 19, 1627 (1978); Rep. Math. Phys. 16, 251 (1979).
17. J. Schwinger, Phys. Rev. 92, 1283 (1953).
18. Yu. A. Gol'fand and E. P. Lichtman, JETP Lett. 13, 323 (1971).
19. J. Wess and B. Zumino, Nucl. Phys. B70, 39 (1974).
20. A. Salam and J. Strathdee, Nucl. Phys. 76B, 477 (1974).
21. J. Wess and B. Zumino, Nucl. Phys. 78B, 1 (1974); S. Ferrara and B. Zumino, Nucl. Phys. B79, 413 (1974); A. Salam and J. Strathdee, Phys. Lett. 51B, 353 (1974).

22. P. A. M. Dirac, Ann. Math. 36, 657 (1935); Proc. R. Soc. A155, 447 (1936); Max-Planck Festschrift, Berlin, 1958 (unpublished).
E. Schrödinger, Proc. R. Irish Acad. A46, 25 (1940). K. Goto, Prog. Theor. Phys. 12, 311 (1954). F. Gürsey, in Istanbul Summer School 1962 (Gordon and Breach, New York, 1963). F. Gürsey, in Relativity, Groups and Topology, Les Houches Summer School 1963 (Gordon and Breach, New York 1963). F. Gürsey and T. D. Lee, Proc. Natl. Acad. Sci. USA 49, 179 (1963). M. Gutzwiller, Helv. Phys. Acta 29, 313 (1956).
W. Thirring, Acta. Phys. Austr. Suppl. 4; O. Nachtmann, Commun. Math. Phys. 6, 1 (1967). G. Börner and H. P. Dürr, Nuovo Cimento 64, 669 (1969) and references cited therein.
23. C. Fronsdal, Rev. Mod. Phys. 37, 221 (1965).
24. C. Fronsdal, Phys. Rev. D10, 589 (1974).
25. J. Fang and C. Fronsdal, Lett. Math. Phys. 2, 391 (1978).
26. C. Fronsdal, Phys. Rev. D20, 848 (1979); J. Fang and C. Fronsdal, Phys. Rev. D22, 1361 (1980).
27. W. Keck, J. Phys. A8, 1819 (1975).
28. C. Fronsdal, Lett. Math. Phys. 1, 165 (1976).
29. M. Kline, Mathematical Thought from Ancient to Modern Times, Oxford University Press, N.Y. 1972. See page 140.
30. Ref. 29, Section 17.4.
31. C. Fronsdal, Phys. Rev. D26, 1988 (1982).

UNITARY REPRESENTATIONS OF SUPERGROUPS

by

C. Fronsdal and T. Hirai

ABSTRACT. To find the obstructions to integrability of representations of superalgebras, we have studied the problem of deciding on a fruitful definition of supergroup. The most natural definition turns out to be a semidirect product, $G = X \cdot G_o$, where G_o is a Lie group and the invariant subsupergroup is in a certain sense nilpotent. Integrability and unitarizability are now precisely defined. It is shown that every irreducible representation of a superalgebra has finite reduction on the even part. For explicit construction of representations we use induction from the even part. It is shown that there exists an invariant measure for integration on supergroups, and this measure is determined. It is shown that, if the inducing representation has a nondegenerate, invariant Hermitean form, then a nondegenerate invariant form on the induced representation is determined explicitly in terms of the invariant measure. The algebras osp(2n/1) are studied in more detail. All unitarizable representations of these superalgebras have a minimal weight.

C. Fronsdal (ed.), Essays on Supersymmetry, 15–66.

0. Introduction

Lie superalgebras are becoming increasingly important, both in physics and in mathematics. Decisive work on the classification problem was done by Kac,[1] who also studied the finite dimensional representations.[2] In physics the infinite dimensional representations are much more interesting; they are the main concern of this paper.

Among the representations of ordinary Lie algebras, two types are especially important: (i) integrable representations and (ii) unitarizable representations. Integrability of a Lie algebra representation is a very exceptional property, yet most physical applications deal exclusively with integrable representations. A unitarizable representation is a special type of integrable representation, of particular importance in the physical applications. The classification of unitarizable representations of Lie algebras is a central problem in mathematics.

In this paper we study the question of integrability and unitarizability of representations of superalgebras. We begin, in Section 1.1, by defining the local supergroup associated with any superalgebra. This turns out to have the structure of a semi-direct product of a local Lie group and a "nilpotent" group, and this discovery leads directly and naturally to the concept of a Lie supergroup associated with any Lie superalgebra. (Theorem 1.1.4 and Definition 1.1.5.) The question of integrability can now be posed precisely, in Section 1.2, and in fact it has become almost trivialized. The obstructions to integrability are essentially those of the Lie subalgebra. (Theorem 1.2.2.)

Unitarizability of an integrable representation of a Lie superalgebra is a two-part problem: (i) construction of a representation of a Lie group by unitary operators in a Hilbert space $\mathcal{H}$ and (ii) verification of selfadjointness of the operators of the odd part of the superalgebra, with respect to the Hermitean form of $\mathcal{H}$ (Definition 1.2.3). A crucial point for our investigation of unitarizability of representations of superalgebras is Proposition 1.2.7 and especially Corollary 1.2.8, according to which every irreducible representation of a supergroup has a finite reduction on the even subgroup. For the actual construction of irreducible representations we use induction from the even subgroup, discussed in Section 1.3. Its generality is assured by Theorem 1.3.4, under a very mild and probably superfluous technical assumption. Induced representations lead to a natural and useful definition of functions on supergroups (Definition 1.3.13).

The central result of this paper is the proof of the existence of a left invariant supermeasure on all supergroups; this is presented as Theorem 2.2.20. It is the result of an investigation of the structure of supergroups in Section 2.1, and a local analysis of the meaning of the left invariant measure on Lie groups in Section 2.2. The measure is defined globally in terms of the left invariant Haar measure on the Lie subgroup.

The existence of an invariant measure on supergroups has the following important consequences (Theorem 1.2.9). Let (π_{in}, V_{in}) be any representation of the Lie subalgebra of a superalgebra, possessing a nondegenerate invariant Hermitean form $<\ ,\ >$, and let (π, V) be the corresponding induced representation of the full superalgebra. Then, schematically,

$$(\ ,\ ) = \int d_\ell \xi <\ ,\ >\ , \quad d_\ell \xi = d\xi^m \ldots d\xi^1\ \omega_1[\xi]\ ,$$

is a nondegenerate invariant Hermitean form for (π, V). The integral is to be understood in the sense of Berezin,[3] over a Grassmann algebra $S_1{}^*$ that is dual to the symmetric algebra S_1 of the odd part of the superalgebra. The factor $\omega_1[\xi]$ is determined by the left invariant measure on the associated supergroup.

It has been conjectured that the verification of unitarizability for representations of superalgebras should be elementary. Here we have confirmed this, by giving a natural and meaningful definition of unitarizability, and by providing a means of verification of unitarizability by a finite algorithm. In fact, the existence of an invariant Hermitean form on induced representations reduces the problem to a Lie group theoretical one, except for the verification of positivity. But in the case of superalgebra representations, induced from the Lie subalgebra, this is a finite, algebraic problem. The details are discussed in Section 2.3.

Some readers may prefer, before approaching the general theory, to have a look at Sections 3.1 and 3.2, where the superalgebra osp(2n/1) is studied in detail. All unitarizable representations are minimal weight representations. We prove that minimal weight representations of osp(2n/1) are unitarizable for sufficiently positive minimal weights.

The existence of invariant integration on supergroups would have made possible the construction of a supergroup analogue of the regular representation of Lie groups, except for the fact that the invariant measure is not positive. We believe that it may be possible to determine large classes of functions on supergroups on which the measure nevertheless determines a positive definite (or semidefinite) norm. Some possibilities are proposed at the end of Section 2.3.

1. General Structural Problems

1.1. The definition of supergroup.

With Kac,[1] we define a superalgebra as a Z_2 graded algebra $\mathfrak{g} = \mathfrak{g}_0 + \mathfrak{g}_1$, and a Lie superalgebra as a superalgebra with an operation [,] satisfying

$$[a,b] = -(-)^{\sigma(a,b)} [b,a]$$

$$[a,[b,c]] = [[a,b],c] + (-)^{\sigma(a,b)} [b,[a,c]] ,$$

where $\sigma(a,b) = \deg(a) \deg(b)$ and

$$\deg(a) = r \leftrightarrow a \in \mathfrak{g}_r , \quad r = 0,1 .$$

An element of $\mathfrak{g}$ is called even (odd) if it belongs to $\mathfrak{g}_0$ ($\mathfrak{g}_1$). In view of a Poincaré-Birkhoff-Witt theorem for Lie superalgebras,[1] we may define the universal enveloping algebra U($\mathfrak{g}$) as the quotient of the tensor algebra over $\mathfrak{g}$ by the ideal generated by the structure relations of $\mathfrak{g}$. If $\{K_\mu\}$ $\mu = 1,\ldots,m$ is a basis for $\mathfrak{g}_1$, and $\{L_j\}$ $j = 1,\ldots,n$ is a basis for $\mathfrak{g}_0$, then the set of monomials

$$\{K_1^{k_1} \ldots K_m^{k_m} L_1^{\ell_1} \ldots L_n^{\ell_n}\} , \quad k_\mu = 0,1, \; \ell_j = 0,1,\ldots, \qquad (1.1.1)$$

forms a basis for U($\mathfrak{g}$).

A definition of formal supergroup has been proposed by Berezin and Kats,[4] and another notion of supergroup is used by physicists.[5] We prefer to define a preliminary concept of local supergroup, in closer analogy with that of local group as used by Dynkin[6] and others. Examination of the structure of local supergroups will lead to a natural definition of supergroup. Let $U(\mathfrak{g}_0)$ be the universal enveloping algebra of the Lie algebra $\mathfrak{g}_0$, and $U'(\mathfrak{g}_0)$ the extension to formal power series. Then the local group associated with $\mathfrak{g}_0$ is the local group of formal exponential power series:

$$e^a = \sum_k \frac{1}{k!} a^k ,$$

with a in $\mathfrak{g}_0$, a^k in $U(\mathfrak{g}_0)$ and e^a in $U'(\mathfrak{g}_0)$. The formula

$$e^a e^b = e^{c(a,b)}$$

defines the Campbell-Hausdorff series

$$c(a,b) = a + b + \frac{1}{2} [a,b] + \dots .$$

This series converges when both a and b are in a certain neighborhood of the origin, but there is (usually) no invariant neighborhood with this property; a local group is not a group.

Attempting a similar construction for superalgebras, one encounters the difficulty that the product of two formal exponentials is not a formal exponential. So far, we have tacitly assumed that the fields over which the algebras are defined are commutative. However, if the field is instead "graded commutative," then the local supergroup can be defined in close analogy with the local group. This circumstance, it seems to us, furnishes adequate motivation for introducing the graded commutative field that is used to good effect by physicists.[7]

Let $\mathcal{B} = \mathcal{B}_0 + \mathcal{B}_1$ denote the infinitely generated Grassmann algebra over C, with unit, $\mathcal{B}_0$ denoting the even part and $\mathcal{B}_1$ the odd part. This is an infinite-dimensional Lie superalgebra with $[a,b] = 0$, hence graded commutative. Let $\mathcal{A} = \mathcal{A}_0 + \mathcal{A}_1$ be the subalgebra of $\mathcal{B}$ that consists of all elements in $\mathcal{B}$ of the form $a = a_b + a_s$ (a_b is the body and a_s is the soul of a) with a_b a complex number and a_s nilpotent. Of course $\mathcal{A}_1 = \mathcal{B}_1$, but $\mathcal{A}_0$ is a proper subalgebra of $\mathcal{B}_0$. Let $\tilde{\mathfrak{g}}$ be the even part of the graded tensor product $\mathcal{A} \otimes \mathfrak{g}$. The elements of $\tilde{\mathfrak{g}}$ have the representation

$$\xi \cdot K + x \cdot L = \sum_{\mu=1}^{m} \xi^\mu K_\mu + \sum_{j=1}^{n} x^j L_j ,$$

$$\xi^\mu \in \mathcal{A}_1 , \quad x^j \in \mathcal{A}_0 ,$$

and right multiplication by $\mathcal{A}$ is defined by

$$K_\mu \xi^\nu = -\xi^\nu K_\mu , \quad K_\mu x^j = x^j K_\mu ,$$

$$L_j \xi^\nu = \xi^\nu L_j , \quad L_j x^k = x^k L_j .$$

Let $U(\tilde{\mathfrak{g}})$ denote the enveloping algebra of $\tilde{\mathfrak{g}}$; it is the even part of the enveloping algebra of $\mathcal{A} \otimes \mathfrak{g}$. If $a \in \mathfrak{g}_0$, then the adjoint action ad(a) of a on $\mathfrak{g}$ extends to a derivation of $U(\mathfrak{g}_0)$ and of $U(\mathfrak{g})$. If $a \in \mathfrak{g}_1$, then ad(a) extends to an anti-derivation of $U(\mathfrak{g})$. But $\tilde{\mathfrak{g}}$ is a Lie algebra, and if $a \in \tilde{\mathfrak{g}}$ then the map defined by

$$U(\tilde{\mathfrak{g}}) \ni A \to [a,A] = aA - Aa \in U(\tilde{\mathfrak{g}})$$

is always a derivation of $U(\tilde{\mathfrak{g}})$. This is the key to the definition of supergroups: derivations generate groups, anti-derivations do not.

<u>Definition 1.1.2</u>. Let $\mathcal{A}$ be as above, and $\mathfrak{g}$ a Lie superalgebra. Then the <u>local supergroup</u> associated with $\mathfrak{g}$ is the local group of formal

exponential power series

$$e^a = \sum_k \frac{1}{k!} a^k , \quad a \in \tilde{\mathfrak{g}} ,$$

where $\tilde{\mathfrak{g}}$ is the even part of $\mathcal{A} \otimes \mathfrak{g}$.

The preceding remarks should make it obvious that this is in fact a local group. In particular, the Campbell-Hausdorff formula holds unchanged.

Every odd element of $\mathcal{A}$ is nilpotent of order 2, therefore any element of $\tilde{\mathfrak{g}}$ of the form

$$\xi \cdot K = \sum_{\mu=1}^{m} \xi^\mu K_\mu , \quad m = \dim(\mathfrak{g}_1) ,$$

is nilpotent of order not greater than m + 1. This fact leads to the structural theorem 1.1.4 below; we first prove

Lemma 1.1.3. For every $a \in \mathfrak{g}$ there is a unique factorization

$$e^a = e^{\xi \cdot K + x \cdot L} e^{y \cdot L} ,$$

$$\xi \cdot K = \sum \xi^\mu K_\mu , \quad x \cdot L = \sum x^j L_j ,$$

with ξ^μ odd, x^j and y^j even, y^j in C and ξ^μ, x^j nilpotent.

Proof. Define α, z, y by

$$a = \alpha \cdot K + z \cdot L + y \cdot L ,$$

with y^j in C and z^j nilpotent. Next, define ξ and x by the Campbell-

Hausdorff formula

$$e^{a}\, e^{-y\cdot L} = e^{\xi\cdot K + x\cdot L} .$$

There remains to prove that ξ and x are nilpotent. Indeed, since α and z are nilpotent,

$$e^{a} = \Sigma \frac{1}{k!} (\alpha\cdot K + z\cdot L + y\cdot L)^{k}$$

$$= A + \Sigma \frac{1}{k!} (y\cdot L)^{k} = (1+B)\, e^{y\cdot L} ,$$

where A and therefore also B is nilpotent. Hence

$$e^{a}\, e^{-y\cdot L} = 1 + B = e^{\xi\cdot K + x\cdot K} ,$$

with B and therefore also ξ and x nilpotent. [Of course, for ξ this is trivial.]

Theorem 1.1.4. The local supergroup associated with $\mathfrak{g} = \mathfrak{g}_0 + \mathfrak{g}_1$ is the semidirect product X·Y, where

$$X = \{e^{a},\ a \in \tilde{\mathfrak{g}} ,\ a \text{ nilpotent}\}$$

is the invariant local sub-supergroup and Y is the local group associated with $\mathfrak{g}_0$. The action of $e^{b} \in Y$ on X is the adjoint action, $e^{a} \rightarrow e^{b} e^{a} e^{-b}$.

Proof. If a,b are both nilpotent, then the Campbell-Hausdorff series terminates, so that c(a,b) is well-defined and nilpotent; hence X is a local supergroup. (The qualification "local" is evidently superfluous in this case.) Next, if $a \in \tilde{\mathfrak{g}}$ is nilpotent, then this property is preserved by the action of Y on X; hence this action leaves X invariant. Finally, Lemma 1.1.3 guarantees the factorization $e^a = AB$, A in X and B in Y.

This simple fact is important, since it allows a straightforward definition of supergroup, without the qualification "local." First of all, as noted parenthetically in the course of the proof of Theorem 1.1.4, we may drop the qualification in the case of the invariant sub-supergroup X.

Definition 1.1.5. Let $\mathfrak{g} = \mathfrak{g}_0 + \mathfrak{g}_1$ be a Lie superalgebra. A supergroup with Lie superalgebra $\mathfrak{g}$ is a semi-direct product

$$G = X \cdot G_0 ,$$

where X was defined in Theorem 1.1.4, G_0 is a Lie group with Lie algebra $\mathfrak{g}_0$, and where the differential of the action of G_0 on X coincides with the adjoint action of $\mathfrak{g}_0$ on X. If G_0 is connected then we say that G is a connected supergroup. If $G' = X \cdot G_0'$, where G_0' is a covering group of G_0, then we say that G' is a covering supergroup of G.

The existence of at least one supergroup with Lie superalgebra $\mathfrak{g}$ is automatic; in particular:

Definition 1.1.6. Let $\mathfrak{g}$ be a Lie superalgebra. The adjoint supergroup of $\mathfrak{g}$ is the supergroup $G = X \cdot G_0$, where G_0 is the connected matrix group defined by the adjoint action of $\mathfrak{g}_0$ on $\mathfrak{g}$. [G_0 is a covering of the adjoint group of $\mathfrak{g}_0$.]

Theorem 1.1.7. If G is a supergroup with Lie superalgebra $\mathfrak{g}$, then G is a covering of the adjoint supergroup of $\mathfrak{g}$.

1.2. Representations of supergroups.

A representation of a superalgebra $\mathfrak{g}$ is a homomorphism (see Section 3.2) onto a superalgebra of linear operators. Finite dimensional representations were investigated by Kac[2] and others.[8] Infinite dimensional representations have received some attention in connection with the physical applications.[9-11]

Every representation of $\mathfrak{g}_o$ (a Lie algebra) extends to a representation of $U(\mathfrak{g}_o)$, but not necessarily to a Lie group. The obstruction is related to problems with the convergence of the exponential series, and ultimately to topological properties of Lie groups. What are the obstructions in the case of superalgebras?

Every representation of $\mathfrak{g}$ (a Lie superalgebra) extends to representations of $\tilde{\mathfrak{g}}$ and of $U(\tilde{\mathfrak{g}})$, hence to a representation of the ideal

$$X(\mathfrak{g}) = \{e^a \ , \ a \in \tilde{\mathfrak{g}} \ , \ a \ \text{nilpotent}\} \ .$$

The following definition is therefore natural.

Definition 1.2.1. A representation of a supergroup $G = X \cdot G_o$ with Lie superalgebra $\mathfrak{g}$ is a triple (π, π_o, V), where V is a complex, topological vector space, π is a representation of $\mathfrak{g}$ in V and π_o an analytic representation of G_o in V, such that (i) the differential of π_o coincides with $\pi|\mathfrak{g}_o$ and (ii) for all $x \in G_o$ and for all $a \in \mathfrak{g}_o$, $\pi_o(x)\, \pi(a)\, \pi_o(x^{-1}) = \pi(Ad_x\, a)$.

If (π, π_o, V) is a representation of a supergroup with Lie algebra $\mathfrak{g}$, then we call the representation (π, V) of $\mathfrak{g}$ its differential. The problem of integrability is now almost trivial.

Theorem 1.2.2. A representation (π,V) of $\mathfrak{g}$ in a complex, topological vector space V is the differential of a representation (π,π_0,V) of a supergroup $G = X\cdot G_0$ if and only if $\pi|\mathfrak{g}_0$ is the differential of an analytic representation of G_0 in V.

Proof. The only point to be checked is that the action of $\mathfrak{g}_0$ on X is the differential of the action of G_0 on X, and even this is obvious since X is a finite dimensional $\mathfrak{g}_0$-module.

A representation of a Lie algebra $\mathfrak{g}_0$ is said to be unitarizable if (i) it is the differential of an analytic representation (π_0,V) of a Lie group G_0 in a complex vector space V and (ii) this representation (π_0,V) is the restriction to the space of analytic vectors of a unitary representation of G_0 in a Hilbert space $\mathcal{H}$. This implies that the operators of the original representation of $\mathfrak{g}_0$ are essentially anti-selfadjoint operators in $\mathcal{H}$, with common dense, invariant domain V. The only new feature that is introduced by superalgebras is that the odd part $\mathfrak{g}_1$ of $\mathfrak{g}$ may be represented by operators that are essentially selfadjoint.

Definition 1.2.3. A representation (π,V) of a superalgebra $\mathfrak{g} = \mathfrak{g}_0 + \mathfrak{g}_1$ is called unitarizable if (i) it is the differential of a representation (π,π_0,V) of a supergroup $G = X\cdot G_0$, (ii) (π_0,V) is the restriction to the space of analytic vectors of a unitary representation of G_0 in a Hilbert space $\mathcal{H}$ and (iii) the operators $\pi(a)$ for $a \in \mathfrak{g}_1$ are essentially selfadjoint operators in $\mathcal{H}$.

In view of Theorem 1.2.2, a representation (π,V) of $\mathfrak{g}$ is unitarizable if and only if (i) its restriction to $\mathfrak{g}_0$ is unitarizable to a unitary representation of a Lie group G_0 in a Hilbert space $\mathcal{H}$ and (ii) the restriction of π to the odd part of $\mathfrak{g}$ is Hermitean with respect to the inner product of $\mathcal{H}$. [Note that this statement does not depend on the existence of a supergroup $G = X \cdot G_o$, for G_o does not necessarily act on X. But there is a covering G_o' of G_o that does act on X, and the relevant supergroup is $G' = X \cdot G_o'$.]

The next result will show that the problem of classifying the representations of $G = X \cdot G_o$ reduces, more or less, to the same problem for G_o. We first introduce a very useful object. If T_1 is the tensor algebra over $\mathfrak{g}_1$, and I_1 is the two-sided ideal generated by all elements of the form $x \otimes y + y \otimes x$, with x,y in $\mathfrak{g}_1$, then T_1/I_1 is the exterior algebra over $\mathfrak{g}_1$. Adding a unit, we get a Grassmann algebra of dimension 2^m, $m = \dim(\mathfrak{g}_1)$.

Definition 1.2.4. The symmetric algebra of $\mathfrak{g}_1$, denoted S_1 or $S(\mathfrak{g}_1)$, is the exterior algebra over $\mathfrak{g}_1$, with unit.

Lemma 1.2.5. As a $\mathfrak{g}_0$-module, S_1 is isomorphic to $U(\mathfrak{g})/I(\mathfrak{g}_0)$, where $I(\mathfrak{g}_0)$ is the left ideal generated by $\mathfrak{g}_0$.

Proof. Let

$$K_{\mu\nu\ldots} = \sum_P [P]\, K_{P(\mu)}\, K_{P(\nu)} \cdots , \tag{1.2.6}$$

where the sum is over all permutations of the indices, [P] is the signature of P, and the number of factors ranges from 0 to $m = \dim(\mathfrak{g}_1)$. [We here interpret K, without an index, as the unit.] The sequence (1.1.1) is a basis for $U(\mathfrak{g})$; therefore, so is

$$\{K_{\mu\nu\ldots} L_1^{\ell_1} \ldots L_n^{\ell_n}\}, \quad \ell_j = 0,1,\ldots .$$

Hence (1.2.6) is a basis for the image of the canonical injection of

$U(\mathfrak{g})/I(\mathfrak{g}_0)$ into $U(\mathfrak{g})$. But this is also a basis for the image of the canonical injection of S_1 into T_1 (with unit), which proves the lemma.

Proposition 1.2.7. Let (π,π_0,V) be an irreducible representation of G, and let V_0 be any G_0-invariant subspace of V. Then the representation (π_0,V) of G_0 is equivalent to a subquotient of $S_1 \otimes V_0$.

Proof. Let $\tau(S_1)$ be the image of the canonical injection of S_1 into $U(\mathfrak{g})$; that is, the complex span of (1.2.6). Since V is irreducible, the space $U(\mathfrak{g})\, V_0 = \tau(S_1)\, V_0$ is dense in V, and as a $\mathfrak{g}_0$-module this is evidently a submodule of $S_1 \otimes V_0$.

A corollary of this result shows that the theory of irreducible representations of supergroups is a relatively elementary extension of the theory of representations of Lie groups.

Corollary 1.2.8. Let (π,π_0,V) be any irreducible representation of G $= X{\cdot}G_0$. Then (π_0,V) is equivalent to the direct sum of a finite set of indecomposable representations of G_0.

Proof. In the contrary case V would be the direct sum of a countable sequence of (not necessarily indecomposable) G_0-invariant subspaces. If V_0 is one of them, then $S_1 \otimes V_0$ cannot be dense in V and V cannot be an irreducible $\mathfrak{g}$-module.

1.3. Induced representations.

Induction is a powerful means of constructing representations of Lie groups. Examples of induced representations of Lie supergroups have been considered by physicists.[9]

Definition 1.3.1. Let $G = X{\cdot}G_o$ be a supergroup, and (π_{in}, V_{in}) an analytic representation of G_o. The space

$$V = U(\mathfrak{g}) \underset{U(\mathfrak{g}_o)}{\otimes} V_{in}\,, \tag{1.3.2}$$

with the action π of $\mathfrak{g}$ induced by left action of $\mathfrak{g}$ on $U(\mathfrak{g})$, is denoted

$$(\pi, V) = \mathrm{IND}\underset{G_o}{\overset{G}{\uparrow}} (\pi_{in}, V_{in}) \tag{1.3.3}$$

and is called the representation of G induced by the representation (π_{in}, V_{in}) of G_o.

How general is this construction? We need a technical, and probably superfluous condition. Let (π', π_o', V') be any irreducible representation of G. We know from Corollary 1.2.8 that (π_o', V') is a direct sum of a finite set of indecomposable representations. We have not been able to exclude the (remote) possibility that each indecomposable subrepresentation may have an infinite composition series; therefore we cannot be sure that V' always contains a G_o-invariant, irreducible subspace.

Theorem 1.3.4. Consider an irreducible representation (π', π_o', V') of G. Assume that V' contains an irreducible, G_o-invariant subspace V_{in}; then (π', π_o', V') is contained as a subquotient in the representation (π, V) of G induced by (π_{in}, V_{in}), with $\pi_{in} = \pi_o'|V_{in}$.

Proof. $U(\mathfrak{g})\, V_{in}$ is dense in V' and equivalent to a subquotient of V.

This justifies our interest in induced representations. As

(π_{in}, V_{in}) runs through all equivalence classes of irreducible representations of G_o, all equivalence classes of irreducible representations of G are obtained (by finite reduction and passage to quotients) from the induced representations; except possibly some pathological representations that have the exotic property of containing no irreducible, G_o-invariant subspace.

Our strategy will be to choose (π_{in}, V_{in}) irreducible and unitarizable, and investigate the unitarizability of the irreducible subquotients of (π, V). The main result will be an explicit expression for an invariant Hermitean form on (π, V). Here we prepare the ground by describing the induced representation in some detail.

We use the basis (1.2.6) for the canonical injection of $U(\mathfrak{g})/I(\mathfrak{g}_o)$ into $U(\mathfrak{g})$. If $\{\psi_a\}$ is a basis for V_{in}, then $\{K_{\mu\nu\ldots} \otimes \psi_a\}$ is a basis for V. This justifies the following.

Definition 1.3.5. Let (π_{in}, V_{in}) be any G_o-module, and (π, V) the corresponding induced representation of G. Any element Φ of V has the representation (finite expansion)

$$\Phi[K] = 1 \otimes \phi + K_\mu \otimes \phi^\mu + K_{\mu\nu} \otimes \phi^{\mu\nu} + \ldots \tag{1.3.6}$$

with ϕ, ϕ^μ, ... in V_{in}. We call this the K-representation of V.

The action of the superalgebra $\mathfrak{g}$ on $\Phi[K]$ is fixed by the action of $\mathfrak{g}_o$ in V_{in} and the structure relations of $\mathfrak{g}$. For example,

$$K_\lambda \Phi[K] = K_\lambda \otimes \phi + K_\lambda K_\mu \otimes \phi^\mu + \ldots$$

$$= 1 \otimes \frac{1}{2}[K_\lambda, K_\mu]\,\phi^\mu + K_\lambda \otimes \phi + K_{\lambda\mu} \otimes \frac{1}{2}\phi^\mu + \ldots, \tag{1.3.7}$$

so that K_λ acts on the coefficients by

$$\phi \to \frac{1}{2}[K_\lambda, K_\mu]\,\phi^\mu\,, \qquad \phi^\mu \to \delta_\lambda{}^\mu\,\phi\,, \dots . \tag{1.3.8}$$

The K-representation is sometimes more convenient than the superfield representation, introduced next.

We recall the meaning of integration on a Grassmann algebra in the sense of Berezin.[3] Every element C of the Grassmann algebra generated by (ξ^μ) has a unique decomposition $C = \xi^1 D + E$, where D and E do not contain ξ^1. Define

$$\int d\xi^1\, C \equiv \partial_1 C \equiv D\,.$$

For multiple integrals we write

$$d\xi = d\xi^m \dots d\xi^1\,, \quad d_\ell\xi = d\xi\,\omega_1[\xi]\,,$$

where $\omega_1[\xi]$ is an element of the Grassmann algebra, with unit constant term, to be chosen for our convenience.

Definition 1.3.9. A V_{in}-superfield is an element of the space $S_1{}^* \otimes V_{in}$, with $S_1{}^*$ the Grassmann algebra generated by (ξ^μ) $\mu = 1,\dots,m = \dim(\mathfrak{g}_1)$. Define $\Phi'[\xi]$ in terms of $\Phi[K]$ by the integral transformation

$$\Phi[K] = \int d_\ell\xi\; e^{\xi\cdot K} \otimes \Phi'[\xi]\,; \tag{1.3.10}$$

then $\Phi'[\xi]$ has the representation (finite expansion)

$$\Phi'[\xi] = \phi' + \xi^\mu\,\phi'_\mu + \xi^\mu\,\xi^\nu\,\phi'_{\mu\nu} + \dots ,$$

with coefficients ϕ', ϕ'_μ, ... in V_{in}. We call this the superfield representation of V. In (1.3.10) the symbol $\otimes$ commutes with $U(\mathfrak{g}_o)$ and with S_1*.

The main purpose of making this transformation, from the K-representation to the superfield representation, is that it will allow us to find an invariant Hermitean form on V, expressed as an integral over S_1*. Since this is the representation that we shall make use of from now on, we define the operators $\pi(a)$ in V by

$$a\,\Phi[K] = \int d_\ell\xi\; e^{\xi\cdot K} \otimes \pi(a)\,\Phi'[\xi]\,, \quad a \in \mathfrak{g}\,. \tag{1.3.11}$$

In the next section we shall study the Hermiticity of these operators. It will be shown that, if (π_{in}, V_{in}) admits an invariant, Hermitean form, then so does (π, V).

The above construction may be applied to Lie groups, although it is only of a formal nature (except in the case of groups with the topology of R^n). Consider, in particular, the case when the inducing subgroup is the unit subgroup, then $\Phi[K]$ is in the enveloping algebra, the ξ^μ are real parameters and $\Phi'[\xi]$ is a function on the group. If $d_\ell\xi$ is the left invariant Haar measure, then the operators $\pi(a)$ are the vector fields of the left translations. Returning to the supergroup G, we can write (formally, unless the topology of G_o is that of R^n)

$$\Phi[K,L] = \int d_\ell\xi\; q(x)\; d_\ell x\; e^{\xi\cdot K}\; e^{x\cdot L} \otimes \Phi'[\xi,x]\,. \tag{1.3.12}$$

Here $\Phi[K,L]$ is in $U(\mathfrak{g})$ and $\Phi'[\xi,x]$ is an S_1*-valued function on G_o. It turns out that, if $d_\ell x$ is the left invariant Haar measure on G_o, and for an appropriate choice of the function $q(x)$ and the measure $d_\ell\xi$, $\mathfrak{g}$ acts on $\Phi'[\xi,x]$ by first order differential operators. Note that, though

(1.3.12) has only a formal or local meaning, because of the global topological nature of G_o, Eqs. (1.3.10) and (1.3.11) are perfectly well defined. We shall make no further use of (1.3.12), except that it suggests the following.

Definition 1.3.13. A function on a supergroup $G = X \cdot G_o$ is an S_1*-valued function on G_o.

Superfields were first used by Salam and Strathdee[7] for the super Poincaré algebra, by Keck[9] for osp(4/1) and by Aneva et al.[9] for su(2,2/1). A geometrical interpretation was developed by Arnowitt and Nath,[12] and an algebraic interpretation was suggested by Fronsdal.[11]

2. Invariant Hermitean Forms

2.1. The invariant vector fields.

Here we prove some general structural identities in the enveloping algebra of a superalgebra $\mathfrak{g} = \mathfrak{g}_o + \mathfrak{g}_1$. They will be used later to evaluate the invariant measure on supergroups and to prove the invariance of a Hermitean form on the space of an induced representation.

As before, let $\{K_\mu\}$ $\mu = 1,\ldots,m$ and $\{L_j\}$ $j = 1,\ldots,n$ be a basis for $\mathfrak{g}_1$ and a basis for $\mathfrak{g}_o$, respectively. The local supergroup consists of elements (choice of "parameterization")

$$F(\xi,x) = e^{\xi\cdot K}\, e^{x\cdot L}\,, \tag{2.1.1}$$

$$\xi\cdot K = \sum_{\mu=1}^{m} \xi^{\mu} K_{\mu}\,, \quad x\cdot L = \sum_{j=1}^{n} x^{j} L_{j}\,,$$

where ξ^{μ} and x^{j} are odd, respectively even, elements of $\mathcal{A}$, the infinitely generated Grassmann algebra over C, with unit. The problem is to determine the quantities ξ', x', $\bar{\xi}$, $\bar{x}$ in

$$e^{\alpha\cdot K} F(\xi,x) = F(\xi',x')\,, \quad e^{y\cdot L} F(\xi,x) = F(\bar{\xi},\bar{x})\,.$$

More precisely, we want to find $\xi'-\xi$, $x'-x$, $\bar{\xi}-\xi$, $\bar{x}-x$ to first order in α and y, for α in $\mathcal{A}_1$ and y in $\mathcal{A}_0$.

The left action of $\mathfrak{g}_0$ is

$$e^{y\cdot L}(e^{\xi\cdot K}\, e^{x\cdot L}) = (e^{y\cdot L}\, e^{\xi\cdot K}\, e^{-y\cdot L})(e^{y\cdot L}\, e^{x\cdot L})$$

$$= e^{\bar{\xi}\cdot K}\, e^{\bar{x}\cdot L}\,. \tag{2.1.2}$$

The first order increment $\bar{\xi}-\xi$ is therefore determined by the adjoint action

$$(\bar{\xi}-\xi)\cdot K = [y\cdot L,\, \xi\cdot K]\,. \tag{2.1.3}$$

To this adjoint action we associate a first order, linear (in ξ^{μ} and in $\partial_{\mu} = \partial/\partial\xi^{\mu}$) differential operator, determined by its action on the linear expression $\xi\cdot K$: $y\cdot L \to -y\cdot s$,

$$-y\cdot s\; \xi\cdot K = [y\cdot L,\, \xi\cdot K]\,. \tag{2.1.4}$$

Thus, if

$$[L_j, K_\mu] = -(M_j)_\mu{}^\nu K_\nu , \tag{2.1.5}$$

then

$$s_j = (M_j)_\mu{}^\nu \xi^\mu \partial_\nu , \tag{2.1.6}$$

the M_j being constant matrices. Thus we have determined $\bar{\xi}$:

$$(\bar{\xi}-\xi)\cdot\partial_\xi = -y\cdot s . \tag{2.1.7}$$

This result can be expressed as the structural identity

$$y\cdot L\, e^{\xi\cdot K} = -y\cdot s\, e^{\xi\cdot K} + e^{\xi\cdot K}\, y\cdot L . \tag{2.1.8}$$

The vector field asociated with the left action of $y\cdot L \in \mathfrak{g}_o$ on G_o will be taken as known, and denoted $y\cdot\kappa$; we would like to relate it to the increment $\bar{x}-x$ by

$$(\bar{x}-x)\cdot\partial_x = -y\cdot\kappa ,$$

in analogy with (2.1.7), but this is possible only as long as $e^{x\cdot L}$ can be interpreted as an element of a Lie group G_o, which is not true here. This is not a real difficulty, however. The factor $e^{x\cdot L}$ plays the role of an element of a $\mathfrak{g}_o$-module and will eventually be replaced by one. The real purpose of this section is to derive the two structural identities (2.1.8) and (2.1.10).

Recall that ξ' and x' were defined by

$$e^{\alpha\cdot K}\, e^{\xi\cdot K}\, e^{x\cdot L} = e^{\xi'\cdot K}\, e^{x'\cdot L} ,$$

and define δx by

$$e^{\delta x\cdot L}\, e^{x\cdot L} = e^{x'\cdot L} .$$

Then (always to first order in α)

$$e^{\alpha\cdot K-\delta x\cdot L}\, e^{\xi\cdot K}\, e^{x\cdot L} = e^{\xi''\cdot K}\, e^{x\cdot L} ,$$

$$\xi''\cdot K = \xi'\cdot K - [\delta x\cdot L, \xi\cdot K] .$$

That is, there is a unique linear function, $\alpha \to \delta x$, such that

$$e^{\alpha\cdot K-\delta x\cdot L}\, e^{\xi\cdot K} = e^{\xi''\cdot K} . \qquad (2.1.9)$$

Hence there is a structural identity of the form

$$\alpha\cdot K\, e^{\xi\cdot K} = [\delta x\cdot L + (\xi''-\xi)\cdot\partial_{\xi}]\, e^{\xi\cdot K} . \qquad (2.1.10)$$

This simple formula determines both δx and $\xi''-\xi$ as linear functions of α. It remains only to find the explicit formulas for δx and $\xi''-\xi$.

If $e^a e^b = e^{b+\delta b}$, then the Campbell-Hausdorff formula, to first order in a, is

$$\delta b = a + \frac{1}{2}[a,b] + \frac{1}{12}[[a,b],b] + \ldots$$
$$= [1 + \frac{1}{2}p + p^2 Q(p^2)]\, a ,$$

where p is the operator

$$pa = [a,b] \,, \tag{2.1.11}$$

and $Q(p^2)$ is the formal power series defined by

$$p^2Q(p^2) = \frac{p}{2}\coth\frac{p}{2} - 1 \,. \tag{2.1.12}$$

Now take--see Eq. (2.1.9)--

$$a = \alpha\cdot K - \delta x\cdot L \,, \quad b = \xi\cdot K \,;$$

then

$$\begin{aligned}\delta b &= (\xi''-\xi)\cdot K \\ &= \alpha\cdot K - \frac{1}{2}[\delta x\cdot L, \xi\cdot K] + p^2Q(p^2)\,\alpha\cdot K \\ &\qquad - \delta x\cdot L + \frac{1}{2}[\alpha\cdot K, \xi\cdot K] - p^2Q(p^2)\,\delta x\cdot L \,. \end{aligned} \tag{2.1.13}$$

The first line is linear in (K_μ) and determines $\xi''-\xi$. The second line is linear in (L_j) and must vanish; this requirement determines δx.

We note that δx is independent of x; Eq. (2.1.10) makes sense independently of the definition of supergroups. There are no problems of globalization here.

So far our calculation has been completely general, but to complete it, we must limit ourselves to a specific family of Lie superalgebras. This will be done later for the family osp(2n/1), but the information contained in (2.1.13) is actually sufficient for our purposes.

We shall also need something about right translations. We have

$$(e^{\xi\cdot K}\, e^{x\cdot L})\, e^{y\cdot L} = e^{\xi\cdot K}\, e^{\bar{x}\cdot L} .$$

Thus, if we define right translation vector fields by

$$e^{\xi\cdot K}\, e^{x\cdot L}\; y\cdot L = y\cdot d\; e^{\xi\cdot K}\, e^{x\cdot L} ,$$

$$y\cdot d = y^j\, d_j{}^\nu\, \partial_\nu + y^j\, d_j{}^k\, \partial_k , \tag{2.1.14}$$

then the $d_j{}^\nu$ vanish identically and the vector fields $d_j{}^k\, \partial_k$ are purely Lie group theoretical. Once again, globalization problems are confined to the Lie subgroup.

Next,

$$e^{\xi\cdot K}\, e^{x\cdot L}\; \alpha\cdot K = e^{\xi\cdot K}\, \alpha^\mu\, (e^{-x\cdot M})_\mu{}^\nu\, K_\nu\, e^{x\cdot L} ,$$

The matrices M_j of the adjoint action were defined in Eq. (2.1.5). Thus if we define the remaining right translation vector fields by

$$e^{\xi\cdot K}\, e^{x\cdot L}\; \alpha\cdot K = \alpha\cdot d\; e^{\xi\cdot K}\, e^{x\cdot L} ,$$

$$\alpha\cdot d = \alpha^\mu\, d_\mu{}^\nu\, \partial_\nu + \alpha^\mu\, d_\mu{}^j\, \partial_j , \tag{2.1.15}$$

then

$$d_\mu{}^\nu = (e^{-x\cdot M})_\mu{}^\lambda\, A_\lambda{}^\nu , \tag{2.1.16}$$

where the matrix A is independent of x.

The property of right translations that will actually be needed is the following. Let us combine the vector fields (2.1.14) and (2.1.15) to

$$D_a^{\ b}\,\partial_b\,, \tag{2.1.17}$$

where a,b run over the combined ranges of $\mu = 1,...,m$ and $j = 1,...,n$. Then the superdeterminant is (since $d_j^{\ \nu} = 0$)

$$|D_a^{\ b}| = \begin{vmatrix} d_\mu^{\ \nu}, & d_j^{\ \nu} \\ d_\mu^{\ k}, & d_j^{\ k} \end{vmatrix} = |d_\mu^{\ \nu}|^{-1}\,|d_j^{\ k}| = |A|^{-1}\, e^{x\cdot \mathrm{tr}\, M}\,|d_j^{\ k}|\,, \tag{2.1.18}$$

with the first factor independent of x.

Finally, we shall derive some consequences of the fact that left translations commute with right translations. According to our two structural identities (2.1.8) and (2.1.10),

$$y\cdot L\, e^{\xi\cdot K}\, e^{x\cdot L} = -(y\cdot s + y\cdot \kappa)\, e^{\xi\cdot K}\, e^{x\cdot L}\,,$$
$$\alpha\cdot\kappa\, e^{\xi\cdot K}\, e^{x\cdot L} = -\alpha^\mu\,(s_\mu^{\ \nu}\,\partial_\nu + i\sigma_\mu^{\ j}\,\kappa_j)\, e^{\xi\cdot K}\, e^{x\cdot L}\,, \tag{2.1.19}$$

where the matrices $(s_\mu^{\ \nu})$ and $(\sigma_\mu^{\ j})$ are independent of x and defined by

$$\alpha^\mu\, s_\mu^{\ \nu}\,\partial_\nu = -(\xi''-\xi)\cdot\partial_\xi + \delta x\cdot s\,,$$
$$\delta x\cdot L = i\,\alpha^\mu\,\sigma_\mu^{\ j}\, L_j\,. \tag{2.1.20}$$

We restrict (x^j) to real values, with $x\cdot L$ within the domain of the exponential bijection between $\mathfrak{g}_0$ and G_0; then (κ^j) can be identified with the vector fields of left translations on G_0. We shall derive some differential equations that are valid in all of G_0.

Since left translations evidently commute with right translations,

$$(a\, e^{\xi\cdot K}\, e^{x\cdot L})\, b = a\, (e^{\xi\cdot K}\, e^{x\cdot L}\, b) ,$$

the vector fields

$$s_j + \kappa_j \quad \text{and} \quad s_\mu{}^\nu\, \partial_\nu + i\, \sigma_\mu{}^j\, \kappa_j$$

of left translations commute with the vector fields of right translations, and in particular with (2.1.15). Since s_j, $s_\mu{}^\nu$ and $\sigma_\mu{}^j$ are independent of x, it follows that they commute with

$$d_\mu{}^\nu\, \partial_\nu = N_\mu{}^\lambda\, A_\lambda{}^\nu\, \partial_\nu , \quad N_\mu{}^\lambda \equiv (e^{-x\cdot M})_\mu{}^\lambda .$$

A short and direct calculation now yields

$$(s_j+\kappa_j)\, \ell n|d_\mu{}^\nu| = \mathrm{tr}\, M_j , \tag{2.1.21}$$

$$(s_\mu + i\sigma_\mu{}^j\, \kappa_j)\, \ell n|d_\lambda{}^\rho| = -\partial_\nu s_\mu{}^\nu . \tag{2.1.22}$$

Globalization will be discussed in Section 2.2.

2.2. Invariant integration on supergroups.

Consider first the case of an ordinary Lie group G. All calculations will first be carried out on a local coordinate domain, leaving globalization a problem to be taken care of subsequently.

Let $x \to (x^a)$ $a = 1,2,\ldots,N$ be a local coordinate system on a neighborhood W of the identity of G, and dx the Lebesque measure. If f is a complex function on G, with support on a subset of W, let

$$\int f \equiv \int dx\; \Omega(x)\; f(x) , \tag{2.2.1}$$

with Ω a fixed function on G. Under left translation by $y \in G$,

$$x \to yx \,, \quad f \to f_y \,, \quad f_y(x) = f(y^{-1}x) \,.$$

The following is a local version of the Haar theorem.

Theorem 2.2.2. There exists a function Ω on W, unique up to a constant factor, such that $\int f_y = \int f$ for all functions f and for all y in W such that f and f_y have support in W.

Proof. A change of variables leads to

$$\int f_y = \int d(yx)\, \Omega(x)\, f(x) \,.$$

The condition on Ω is therefore

$$\Omega(x)/\Omega(yx) = \left|\frac{\partial(yx)}{\partial x}\right| \,. \tag{2.2.3}$$

The right hand side is the Jacobian of the transformation $x^a \to y^a(x) = (yx)^a$. For x,y,z in G, consider the composite mapping $z \to xz \to yxz$; then

$$\left|\frac{\partial(yxz)}{\partial z}\right| = \left|\frac{\partial(yx)}{\partial x}\right| \left|\frac{\partial(xz)}{\partial z}\right| \,.$$

The condition (2.2.3) is thus satisfied by

$$\Omega(x) = \left|\frac{\partial(xz)}{\partial z}\right|^{-1} \,, \tag{2.2.4}$$

for any fixed z. We shall choose z = 1.

It will be useful to consider the differential form of the condition (2.2.3). The differential of the mapping $x \to yx$ at $y = 1$ defines a linear map $\delta y \to \delta y^{(\ell)}$ of the tangent space of G at the identity into itself. In the local coordinate basis this determines a matrix valued function S(x):

$$\delta y^{(\ell)} = (\delta y)^a S_a{}^b \partial_b . \tag{2.2.5}$$

The differential form of (2.2.3) is

$$S_a{}^b \partial_b \ln \Omega + \partial_b S_a{}^b = 0 . \tag{2.2.6}$$

It is not immediately obvious that (2.2.6) has a solution. We now evaluate the function (2.2.4) explicitly and then verify that it satifies Eq. (2.2.6). Under right translation by z in G,

$$x \to xz , \ f \to f^z , \ f^z(x) = f(xz) . \tag{2.2.7}$$

The differential of this mapping at $z = 1$ defines a linear mapping $\delta z \to \delta z^{(r)}$ and a matrix valued function D(x):

$$\delta z^{(r)} = (\delta z)^a D_a{}^b \partial_b . \tag{2.2.8}$$

Eq. (2.2.4) now reduces to, when $z = 1$,

$$\Omega(x) = |D_a{}^b|^{-1} . \tag{2.2.9}$$

To verify (2.2.6) we use the fact that left translations commute with right translations:

$$S_a{}^b \partial_b D_c{}^d = D_c{}^b \partial_b S_a{}^d .$$

Multiplying by $(D^{-1})_d{}^c$ and summing, we get

$$S_a{}^b \partial_b \ell n|D| = \partial_b S_a{}^b , \tag{2.2.10}$$

which in view of (2.2.9) is precisely Eq. (2.2.6).

We now return to the case when G is a Lie supergroup. We proceed heuristically, and then prove a posteriori that the result obtained by this means is indeed correct.

Consider functions on G_o taking values in $S_1{}^*$, and replace the coordinates

$$(x^1,\ldots,x^N) \rightarrow (\xi^1,\ldots,\xi^m, x^1,\ldots,x^n) ,$$

where (x^j) are local coordinates for G_o and (ξ^μ) generate the complex Grassmann algebra $S_1{}^*$. We confine ourselves, temporarily, to the domain of the exponential bijection between $\mathfrak{g}_o$ and G_o; this allows us to use the parameterization of G that was introduced in Section 2.1, with x^j real from now on. The matrix D associated with right translations is just the matrix that was introduced in Eq. (2.1.17), and whose superdeterminant is given by Eq. (2.1.18). Hence we take

$$\Omega(\xi,x) = \omega_1[\xi]\, e^{-x\cdot \mathrm{Tr}\, M}\, \omega_o(x) , \tag{2.2.11}$$

$$\omega_1[\xi] = |A_\mu{}^\nu| , \quad \omega_o(x) = |d_j{}^k|^{-1} . \tag{2.2.12}$$

The conjectured invariant measure on G is thus

$$d_\ell \xi\, e^{-x\cdot \mathrm{tr}\, M}\, \omega_o(x)\, dx , \quad d_\ell \xi = d\xi\, \omega_1[\xi] . \tag{2.2.13}$$

We must now give this a global sense.

The first factor in (2.2.13) needs no globalization, for $\omega_1[\xi]$ is constant on G_o and is well defined.

The factor $\omega_o(x)\,dx$ coincides, within the domain of the exponential map, with the left invariant Haar measure $d_\ell x$ on G_o. We therefore replace

$$\omega_o(x)\,dx \to d_\ell x\,. \tag{2.2.14}$$

The factor $e^{-x\cdot \mathrm{tr}\, M}$ is the determinant of the adjoint action by $e^{-x\cdot L}$ on ω_1. We have coordinatized $x \in G_o$, within the domain of the exponential map, identifying $x = e^{x\cdot L}$; hence $e^{-x\cdot \mathrm{tr}\, M}$ is the expression, in local coordinates, for the function

$$q(x) = \det(\mathrm{Ad}_x|\mathfrak{g}_1)^{-1}\,, \quad x \in G_o\,. \tag{2.2.15}$$

Hence (2.2.13) leads to the following global measure on G:

$$d_\ell \xi\; q(x)\; d_\ell x\,. \tag{2.2.16}$$

We can now globalize the differential equations (2.1.21) and (2.1.22). They are the local expressions for

$$(s_j + \kappa_j)\, \ell n\{\omega_1[\xi]\, q(x)\} = \mathrm{tr}\, M_j\,, \tag{2.2.17}$$

$$(s_\mu + i\sigma_\mu{}^j\, \kappa_j)\, \ell n\{\omega_1[\xi]\, q(x)\} = -\partial_\nu s_\mu{}^\nu\,. \tag{2.2.18}$$

These equations will be used to demonstrate the invariance of the measure (2.2.16).

Consider a space V of functions on the supergroup $G = X{\cdot}G_o$; that is, functions on G_o with values in $S_1{}^*$, with the action of $\mathfrak{g}$ given by the vector fields of the left translations, namely, from Eq. (2.1.19),

$$\begin{aligned} \pi(L_j) &= s_j + \kappa_j , \\ \pi(K_\mu) &= s_\mu + i\, \sigma_\mu{}^j\, \kappa_j . \end{aligned} \tag{2.2.19}$$

Later we shall show that this is in fact an induced representation of G. We are finally in a position to state and to prove one of our main results, the invariance of the measure (2.2.16).

Theorem 2.2.20. (Invariant integration on supergroups.) The operators (2.2.19) satisfy

$$\int d_\ell \xi \; q(x)\; d_\ell x \; \pi(a)\; \Phi'[\xi] = 0 , \tag{2.2.21}$$

for all $a \in \mathfrak{g}$ and for $\Phi'[\xi]$ any differentiable function on G.

Proof. This follows easily from the fact that

$$\int d\xi\; \partial_\mu \Psi'[\xi] = 0 , \quad \int d_\ell x\; \kappa_j\; \Psi'[\xi] = 0 ,$$

with the help of Eqs. (2.2.17) and (2.2.18).

Remark 2.2.22. For x near the identity of G_o we have

$$\kappa_j\; \ell n\, q(x) = \frac{d}{dt}\, \ell n |e^{-x\cdot M + tM_j}|_{t=0} = \mathrm{tr}\, M_j ;$$

hence

$$\kappa_j \, \ell n \, q(x) = \text{tr } M_j \; ; \tag{2.2.23}$$

holds on G_0, and Eqs. (2.2.17) and (2.2.18) reduce to

$$s_j \, \omega_1[\xi] = 0 \tag{2.2.24}$$

$$s_\mu \, \ell n \, \omega_1[\xi] = -i\sigma_\mu{}^j \text{ tr } M_j - \partial_\nu s_\mu{}^\nu \; . \tag{2.2.25}$$

These are differential equations in $S_1{}^*$; all dependence on x having disappeared.

Some examples of invariant measures on supergroups have been known for some time; see Aneva et al.[9] A formal argument for the existence of an invariant measure on supergroups was given by De Witt.[15]

2.3. Invariant Hermitean forms.

We now return to the discussion of Section 1.3. The formulas obtained in Section 2.1 will give us the action π of induced representations, and the results of Section 2.2 will enable us to derive an explicit formula for an invariant Hermitean form.

Let us first of all define operators (κ_j) $j = 1,\ldots,n$ in V_{in} by setting

$$\pi_{in}(L_j) = \kappa_j + \text{tr } M_j \; . \tag{2.3.1}$$

This will allow us to recover the case when V_{in} is a space of functions on G_0 by simply interpreting (κ_j) as the vector fields on left translations.

The operators $\pi(a)$ of the induced representation are defined by (1.3.11). Using the first structural identity, Eq. (2.1.8), we get

$$L_j\Phi[K] = \int d_\ell\xi \; L_j \; e^{\xi\cdot K} \otimes \Phi'[\xi]$$

$$= \int d_\ell\xi \; (-s_j \; e^{\xi\cdot K} + e^{\xi\cdot K} \; L_j) \otimes \Phi'[\xi]$$

$$= \int d_\ell\xi \; e^{\xi\cdot K} \otimes (-\partial_\mu s_j^{\;\mu} + \kappa_j + \mathrm{tr}\, M_j) \; \Phi'[\xi] \; .$$

We used (2.2.24); also $s_j^{\;\mu} = (M_j)_\nu^{\;\mu} \xi^\nu$, so finally

$$\pi(L_j) = s_j + \kappa_j \; . \tag{2.3.2}$$

Similarly, with the help of (2.1.10), and the definitions (2.1.20),

$$K_\mu\Phi[K] = \int d_\ell\xi \; K_\mu \; e^{\xi\cdot K} \otimes \Phi'[\xi]$$

$$= \int d_\ell\xi \; (-s_\mu \; e^{\xi\cdot K} + e^{\xi\cdot K} \; i\sigma_\mu^{\;j} \; L_j) \otimes \Phi'[\xi]$$

$$= \int d_\ell\xi \; e^{\xi\cdot K} \otimes \left[\frac{1}{\omega_1} \partial_\nu s_\mu^{\;\nu} \; \omega_1 + i\sigma_\mu^{\;j} \; (\kappa_j + \mathrm{tr}\, M_j)\right] \Phi'[\xi] \; .$$

Now using (2.2.25) we obtain

$$\pi(K_\mu) = s_\mu + i\sigma_\mu^{\;j} \; \kappa_j \; . \tag{2.3.3}$$

We have thus recovered (2.2.19) and verified that this representation is induced. It is induced by the representation (2.3.1) of $\mathfrak{g}_o$, not only when the κ_j stand for the left translation vector fields on G_o as was the case in the preceding section, but more generally when, as here, $L_j \to \kappa_j$ is any representation of $\mathfrak{g}_o$.

We now confront the problem of whether the existence of an invariant Hermitean form in (π_{in}, V_{in}) induces the same for (π, V).

Actually, this is the right question only if tr $M_j = 0$. Consider first the case when V_{in} is the space of functions f on G_o having finite norm with respect to the inner product

$$<f_1,f_2> = \int q(x)\, d_\ell x\, f_1^*(x)\, f_2(x)\ . \tag{2.3.4}$$

We here suppose that tr M_j is real and that $q(x) > 0$. If (κ_j) are the vector fields of left translations on G, then we have by virtue of Eq. (2.2.23),

$$<f_1,\kappa_j f_2> + <\kappa_j f_1,f_2> + \mathrm{tr}\, M_j <f_1,f_2> = 0\ . \tag{2.3.5}$$

More generally we suppose, from now on, only that (π_{in},V_{in}) is equipped with a Hermitean form < , > such that this equation holds.

It will be assumed, from this point onwards, that $\mathfrak{g}$ is the real superalgebra over the basis (K_μ,L_j). More precisely, we suppose that the structure is expressed by

$$[K_\mu,L_j] = M_{j\mu}{}^\nu\, K_\nu\ ,$$

$$[L_j,L_k] = N_{kj}{}^\ell\, L_\ell\ , \tag{2.3.6}$$

$$\frac{1}{i}[K_\mu,K_\nu] = O_{\mu\nu}{}^j\, L_j\ ,$$

with real coefficients $M_{j\mu}{}^\nu$, $N_{kj}{}^\ell$, $O_{\mu\nu}{}^j$. It follows that S_1 and S_1^* are real $\mathfrak{g}_o$-modules. More precisely, the action of $\mathfrak{g}_o$ in S_1^* commutes with the involutive anti-automorphism $z \to z^\dagger$ generated by $\xi^\mu \to \xi^\mu$, $i \to -i$, with respect to which ξ^μ is real and

$$(\xi^{\mu_1} \ldots \xi^{\mu_k})^\dagger = (-)^{k(k-1)/2}\, \xi^{\mu_1} \ldots \xi^{\mu_k}\ . \tag{2.3.7}$$

The matrices $(\sigma_\mu{}^j)$ and $(s_\mu{}^\nu)$ are real:

$$(\sigma_\mu{}^j)^\dagger = \sigma_\mu{}^j\,,\quad (s_\mu{}^\nu)^\dagger = s_\mu{}^\nu\,. \tag{2.3.8}$$

The inner product < , > of V_{in} will be extended to $S_1{}^* \otimes V_{in}$ by posing

$$<z_1\phi_1, z_2\phi_2> = z_1{}^\dagger z_2 <\phi_1, \phi_2>\,,$$

for z_1, z_2 in $S_1{}^*$. Finally, it is easy to verify that $\omega_1[\xi]$ is a real and even element of the complex Grassmann algebra, with unit constant term.

Theorem 2.3.9. Let $\mathfrak{g}$ be a real superalgebra, and suppose that (π_{in}, V_{in}) has a Hermitean form < , > satisfying (2.3.5), with $\pi_{in}(L_j) = \kappa_j + \mathrm{tr}\, M_j$. Then

$$(\Phi_1, \Phi_2) \equiv \int d_\ell\xi <\Phi_1'[\xi], \Phi_2'[\xi]> \tag{2.3.10}$$

defines an invariant Hermitean form for the induced representation (π,V), in the sense that

$$(\pi(L_j)\,\Phi_1, \Phi_2) + (\Phi_1, \pi(L_j)\,\Phi_2) = 0 \tag{2.3.11}$$

$$(\pi(K_\mu)\,\Phi_1, \Phi_2) - (\Phi_1, \pi(K_\mu)\,\Phi_2) = 0\,. \tag{2.3.12}$$

Proof. Note first that

$$(\partial_\mu\Phi_1)^\dagger\,\Phi_2 = \Phi_1{}^\dagger\,\overleftarrow{\partial}_\mu\,\Phi_2 = \mp(\partial_\mu\Phi_1{}^\dagger)\,\Phi_2$$

$$= \mp\partial_\mu(\Phi_1^\dagger\Phi_2) + \Phi_1^\dagger\,\partial_\mu\,\Phi_2\ ,$$

with upper (lower) sign if Φ_1 is even (odd). Hence ordinary rules for integration by parts apply with a sign change. Now $d_\ell\xi = d\xi\ \omega_1[\xi]$, and Eq. (2.2.24) gives

$$(\Phi_1, s_j\Phi_2) + (s_j\Phi_1, \Phi_2) - \mathrm{tr}\ M_j(\Phi_1, \Phi_2) = 0\ . \tag{2.3.13}$$

Eq. (2.3.11) now follows from (2.3.2), (2.3.5) and (2.3.13), and (2.3.12) in the same way using (2.2.25).

If tr $M_j = 0$, then κ_j is anti-selfadjoint in the metric $<\ ,\ >$, and we may specialize to the case when (π_{in}, V_{in}) is unitarizable in a Hilbert space $\mathcal{H}$, supposing also that $<\ ,\ >$ is the restriction of the Hilbert metric to V_{in}. The tr M_j-term in (2.3.5), if different from zero, prevents this, but not in any essential way.

Definition 2.3.14. Let $G = X{\cdot}G_o$ be a supergroup with Lie algebra $\mathfrak{g} = \mathfrak{g}_o + \mathfrak{g}_1$, and let $x \to q(x)$ be the function on G_o defined by Eq. (2.2.15). A representation $x \to T_x$ of G_o in a Hilbert spaçe with metric $<\ ,\ >$ will be called quasi-unitary if

$$<T_xf_1, T_xf_2> = <f_1, f_2>/q(x)\ . \tag{2.3.15}$$

A representation of $\mathfrak{g}_o$ will be called quasi-unitarizable if it is the differential of a quasi-unitary representation of G_o. [Here G_o is any Lie group such that $G = X{\cdot}G_o$ is a supergroup with superalgebra $\mathfrak{g}$.]

It is clear that (2.3.5) is a necessary condition for the representation (2.3.1) to be quasi-unitarizable. We suppose from now on that (π_{in}, V_{in}) is quasi-unitarizable. It implies, in particular, that

< , > is positive definite and that (π_{in}, V_{in}) is integrable. It follows that (π, V) is also integrable. [Suppose that $G = X \cdot G_o$ and that (π_{in}, V_{in}) is the differential of a representation of G_o. Then (π, V) is the differential of a covering of G.] If the Hermitean form (,) would have been positive, then unitarizability of (π, V) would have followed--but (,) is seldom positive.

The construction of the invariant form (2.3.10) is nevertheless a major step towards the solution of the problem, and of immediate utility. Here are some of the possibilities:

(i) We believe that, with the help of a generalization of the Dirac operator, the induced representation (π, V) can be split into a sum of two similar $\mathfrak{g}$-modules and that, under some conditions on (π_{in}, V_{in}), the invariant Hermitean form can be shown to be positive definite (or negative definite) on one of them. [In some simple examples, where this conjecture was confirmed, the Dirac operator is essentially the second order super Casimir operator.[13]]

(ii) The reductions of (π, V), into sums of indecomposable representations of G or of G_o, are both discrete and finite. Let V^1 be a G-irreducible subspace of V. A necessary condition for (π, V^1) to be unitarizable is that its restriction to G_o be completely reducible,

$$V^1 = V_1{}^1 \oplus \ldots \oplus V_r{}^1 ,$$

with each summand a unitarizable, irreducible representation of G_o. If this test is passed we examine the restriction $(\, , \,)_1$ of the invariant Hermitean form (,) to V^1. A sufficient condition for unitarizability of (π, V^1) is that $(\, , \,)_1$ be positive definite (or negative definite). But the verification of this is a finite problem, for it is enough to select a vector $v_k \neq 0$ from each $V_k{}^1$ and check that $(v_k, v_k) > 0$ for $k = 1,\ldots,r$ (or < 0 for $k = 1,\ldots,r$).

(iii) If $\pi(K_\mu)$, $\mu = 1,\ldots,m$ are selfadjoint, and $(c^\mu) \in R^m$, then $[\pi(c\cdot\kappa), \pi(c\cdot\kappa)] \in i\pi(\mathfrak{g}_o)$ is a positive semidefinite operator. For this reason many $\mathfrak{g}$-modules are minimal weight representations. One expects unitarizability for sufficiently positive minimal weights, and the existence of an invariant inner product allows a simple proof of this, as demonstrated for osp(2n/1) in Section 3.4.

Finally, the existence of a non-degenerate, invariant form on induced representations has an important application in massless quantum field theories.[14] Here the unitary representations are invariably imbedded in nondecomposable representations. Quantization depends on the existence of a nondegenerate symplectic structure, the existence of which is guaranteed for induced representations. In fact, the imaginary part of the invariant Hermitean form is an invariant symplectic form. This is why superfield representations always include the necessary "auxiliary fields." Note, however, that all this applies to induction from $\mathfrak{g}_o$. It is false for some superfields associated with induction from sub-superalgebras. For an example, see the next paper, Section 6.

3. An Example: osp(2n/1)

3.1. Phase space, sp(2n,R) and osp(2n/1).

Phase space, denoted V_{2n}, is a real vector space of dimension 2n, endowed with an antisymmetric bilinear form η. The natural coordinates

$$q_1,\ldots,q_n,p^1,\ldots,p^n = q_1,\ldots,q_{2n}$$

form a basis for the real vector space dual V_{2n}^*. The Poisson bracket

$$\{q_\mu,q_\nu\} = -\eta_{\mu\nu}\,,$$
$$\eta_{\mu\nu} = \pm 1\,,\ \mu = \nu \pm n\,, \tag{3.1.1}$$

$\eta_{\mu\nu} = 0$ otherwise, gives rise to the following structures:

(i) A two-form on V_{2n}^* is defined by

$$\eta(u,v) = \{u,v\} = \eta_{\mu\nu}u^\mu v^\nu\ ;\ \ u,v \in V_{2n}^*\,.$$

(ii) A bijection $V_{2n} \leftrightarrow V_{2n}^*$ is given by

$$\eta(u,v) = v(\tilde{u}) = -u(\tilde{v})\ ;\ \ \tilde{u},\tilde{v} \in V_{2n}\,.$$

This amounts to a rule for raising and lowering indices:

$$\tilde{u}_\mu = u^\nu\,\eta_{\nu\mu}\,,\ \ u^\mu = \eta^{\mu\nu}\,\tilde{u}_\nu\,,\ \ \eta^{\mu\nu} = \eta_{\mu\nu}\,.$$

The tilde will be omitted as the position of the index will convey the same information.

(iii) A family of derivations of $C^\infty(V_{2n})$ is defined by

$$g \to \{f,g\} = -\eta_{\mu\nu} \frac{\partial f}{\partial q_\mu} \frac{\partial g}{\partial q_\nu}, \quad f,g \in C^\infty(V_{2n}) . \tag{3.1.2}$$

To $f \in C^\infty(V_{2n})$ is thus associated a vector field

$$f^\# = -\eta_{\mu\nu} \frac{\partial f}{\partial q_\mu} \frac{\partial}{\partial q_\nu} .$$

The Poisson bracket turns $C^\infty(V_{2n})$ into a Lie algebra. Our interest focuses on the subalgebra

$$sp(2n,R) = \{f = c^{\mu\nu} q_\mu q_\nu, \ c^{\mu\nu} \text{ real}\} ,$$

spanned by

$$f_{\mu\nu} = f_{\nu\mu} = q_\mu q_\nu \ ; \ \mu,\nu = 1,\dots,2n .$$

Here is one of the most fundamental superalgebras:

Definition 3.1.3. The graded Lie algebra (Lie superalgebra) osp(2n/1) is the real vector space spanned by $(f_{\mu\nu}, q_\mu)$, with the following structure relations

$$\{f_{\mu\nu}, f_{\lambda\rho}\}_- = -\eta_{\mu\lambda} f_{\nu\rho} - \eta_{\mu\rho} f_{\nu\lambda} - \eta_{\nu\lambda} f_{\mu\rho} - \eta_{\nu\rho} f_{\mu\lambda} ,$$

$$\{f_{\mu\nu}, q_\lambda\}_- = -\eta_{\mu\lambda} q_\nu - \eta_{\nu\lambda} q_\mu , \tag{3.1.4}$$

$$\{q_\mu q_\nu\}_+ = 2f_{\mu\nu} \quad (= 2iL_{\mu\nu}) .$$

The first two brackets are antisymmetric and derived from (3.1.2), so $\{\ ,\ \}_-$ is just the Poisson bracket. The third bracket is symmetric and coincides with two times the ordinary product of functions. The generators $(f_{\mu\nu})$ span the even part sp(2n,R) of osp(2n/1) and the generators (q_μ) span the odd part. We write $\mathfrak{g}$ for osp(2n/1), $\mathfrak{g}_0$ for the even subalgebra and $\mathfrak{g}_1$ for the odd part; $\mathfrak{g} = \mathfrak{g}_0 + \mathfrak{g}_1$ is a superalgebra in the sense of the definition given in Section 1.1.

The dimensions of $\mathfrak{g}_1$ and $\mathfrak{g}_0$, that we have denoted m and n in dealing with the general case, are now 2n and n(2n+1). The latin index is replaced by a symmetric pair of Greek indices. All Greek indices run from 1 to 2n.

3.2. The oscillator representation.

The oscillator representation is the most singular representation of osp(2n/1), and an important example of a unitarizable representation of a Lie superalgebra. It is defined by the Weyl quantization map.

The Heisenberg algebra is the real vector space of functions on V_{2n} spanned by $q_1,\dots,q_{2n}$ and 1, with the structure of Lie algebra given by the Poisson bracket. This algebra has only one irreducible, faithful representation (up to projective equivalence) by self-adjoint operators in a Hilbert space, and only one irreducible representation (up to unitary equivalence) by selfadjoint operators in a Hilbert space such that

$$q_\mu \to \hat{q}_\mu\,, \quad 1 \to \text{identity operator}\,. \tag{3.2.1}$$

Weyl quantization is an extension of this map, defined by

$$q_\mu \cdots q_\omega \to \text{Symm.}\ \hat{q}_\mu \cdots \hat{q}_\omega\,,$$

where complete symmetrization is to be carried out on the indices.

The oscillator representation is obtained by restriction of the Weyl quantization map to polynomials of order 0,1 and 2. Thus

$$q_\mu \to \hat{q}_\mu \, , \quad f_{\mu\nu} \to \hat{f}_{\mu\nu} = \frac{1}{2}(\hat{q}_\mu \hat{q}_\nu + \hat{q}_\nu \hat{q}_\mu) \, . \tag{3.2.2}$$

These operators are self-adjoint and form a representation of osp(2n/1) in the sense that the structure (3.1.4) is preserved when $q_\mu, f_{\mu\nu}$ are replaced by the operators $\hat{q}_\mu, \hat{f}_{\mu\nu}$ and

$$\{\, , \,\}_- \to \frac{1}{i}[\, , \,]_- \, , \quad \{\, , \,\}_+ \to [\, , \,]_+ \, , \tag{3.2.3}$$

where $[\, , \,]_-$ is the commutator and $[\, , \,]_+$ is the anticommutator. Although a selfadjoint basis is preferred by physicists, the alternative of a basis of anti-selfadjoint operators for Lie algebras (strongly favored by mathematicians) turned out to be much more convenient for our purposes. To accomplish this we must define a representation as a map that preserves the structure when, instead of (3.2.3),

$$\{\, , \,\}_- \to [\, , \,] \, , \quad \{\, , \,\}_+ \to \frac{1}{i}[\, , \,]_+ \, .$$

To facilitate comparison with the other sections of this paper we shall also change the notation slightly. We henceforth write

$$K_\mu \text{ instead of } q_\mu \, , \quad iL_{\mu\nu} \text{ instead of } f_{\mu\nu} \, .$$

In the oscillator representation:

$$K_\mu \to \hat{q}_\mu \, , \quad L_{\mu\nu} \to \frac{1}{i}\hat{f}_{\mu\nu} \, .$$

3.3. Evaluation of the left translation vector fields.

We shall here complete the evaluation that was interrupted at the end of Section 2.1, in the special case when $\mathfrak{g} = osp(2n/1)$. The only change in notation is that $y{\cdot}L = y^j L_j$ is replaced by

$$y{\cdot}L = \frac{1}{2} y^{\mu\nu} L_{\mu\nu} ,$$

a symmetric pair of Greek indices replacing a single latin index.

The structure relations (3.1.4) give

$$[y{\cdot}L, \xi{\cdot}K] = \frac{1}{2} y^{\mu\nu} (\xi_\mu K_\nu + \xi_\nu K_\mu) ,$$

and the definition (2.1.4) gives the following form for $s_{\mu\nu}$:

$$s_{\mu\nu} = - \xi_\mu \partial_\nu - \xi_\nu \partial_\mu . \tag{3.3.1}$$

Next, δx will be calculated from the vanishing of the even part of δb, the second line in Eq. (2.1.13). The structure relations give

$$\frac{1}{2} [\alpha{\cdot}K, \xi{\cdot}K] = -i\alpha^\mu \xi^\nu L_{\mu\nu} ,$$

$$[\delta x{\cdot}L, \xi{\cdot}K] = \delta x^{\mu\nu} \xi_\mu K_\nu \quad .$$

The operator p was defined by Eq. (2.1.11); here $b = \xi{\cdot}K$ and

$$p^2(-\delta x{\cdot}L) = 2i\delta x^{\mu\nu} \xi_\mu \xi^\lambda L_{\nu\lambda} .$$

Iteration gives

$$Q(p^2)\, p^2 (-\delta x{\cdot}L) = 2iQ(-2i\xi^2)\, \delta x^{\mu\nu} \xi_\mu \xi^\lambda L_{\nu\lambda} .$$

[It is this step that we are unable to do in the general case, though it is simple enough in the case at hand.] The condition that the even part of δb vanish is thus

$$\delta x\cdot L + i\alpha^{\mu}\xi^{\nu} L_{\mu\nu} = 2iQ(-2i\xi^2)\, \delta x^{\mu\nu}\, \xi_{\mu}\xi^{\lambda} L_{\nu\lambda}\,. \tag{3.3.2}$$

To solve for δx we note that the map $\alpha \to \delta x$ is $\mathfrak{g}_0$-invariant; this means that there exists a polynomial $g^{-1}(\xi^2)$ such that

$$\delta x\cdot L = -ig^{-1}(\xi^2)\, \alpha^{\mu}\xi^{\nu} L_{\mu\nu}\,, \tag{3.3.3}$$

where ξ^2 is the $\mathfrak{g}_0$-invariant $\xi^{\mu}\,\xi_{\mu}$. Now Eq. (3.3.2) reduces to

$$g(\xi^2) = 1 - 2i\xi^2\, Q(-2i\xi^2)\,, \tag{3.3.4}$$

and δx is thereby determined.

To evaluate the remaining, odd part of δb we first obtain

$$p^2\, \alpha\cdot K = 2i(\alpha\cdot\xi\ \xi\cdot K - \xi^2\, \alpha\cdot K)\,,$$

$$Q(p^2)\, p^2\, \alpha\cdot K = 2iQ(-2i\xi^2)(\alpha\cdot\xi\ \xi\cdot K - \xi^2\, \alpha\cdot K)\,.$$

The expression (2.1.13) for δb now reduces to

$$(\xi''-\xi)\cdot K = g(\xi^2)(1 + i\xi^2/2g^2)\, \alpha\cdot K$$

$$+ (1/\xi^2)[1 - g(\xi^2)(1 + i\xi^2/2g^2)]\, \alpha\cdot\xi\ \xi\cdot K\,.$$

This may be marvellously simplified by a change of variables and by making use of special properties of the function $g(\xi^2)$. Let

$$\theta^\mu = -g^{-1}(\xi^2)\,\xi^\mu \,. \tag{3.3.5}$$

The transformation $\xi^\mu \to \theta^\mu$ is an isomorphism of Grassmann algebras. The inverse relation

$$\xi^\mu = -h(\theta^2)\,\theta^\mu \tag{3.3.6}$$

defines the polynomial $h(\theta^2)$. Now

$$(\xi''-\xi)\cdot K = (1 + i\theta^2/2)\,h(\theta^2)\,\alpha\cdot K$$

$$+ (1/\theta^2)[1 - (1 + i\theta^2/2)\,h(\theta^2)]\,\alpha\cdot\theta\,\theta\cdot K \,.$$

Using (2.1.2) and (3.3.4) we find the differential equation

$$2\theta^2\,(1 + i\theta^2/2)\,h'(\theta^2) = 1 - (1 + i\theta^2/2)\,h(\theta^2)\,,$$

where h' is the derivative of h, so that

$$(\xi''-\xi)\cdot K = (1 + i\theta^2/2)[h(\theta^2)\,\alpha\cdot K + 2h'(\theta^2)\,\alpha\cdot\theta\,\theta\cdot K]$$

$$= -(1 + i\theta^2/2)\,\alpha\cdot\partial_\theta\,\xi\cdot K \,. \tag{3.3.7}$$

The structural identity (2.1.10) becomes

$$\alpha\cdot K\, e^{\xi\cdot K} = [i\alpha^\mu\theta^\nu L_{\mu\nu} - (1 + i\theta^2/2)\,\alpha\cdot\partial_\theta]\, e^{\xi\cdot K}\,,$$

and the operators (2.2.19), first introduced locally in Eqs. (2.1.19) and (2.1.20),

$$\pi(L_{\mu\nu}) = s_{\mu\nu} + \kappa_{\mu\nu}\,, \qquad s_{\mu\nu} = -\,\theta_\mu\partial_\nu - \theta_\nu\partial_\mu\,, \tag{3.3.8}$$

$$\pi(K_\mu) = s_\mu + i\sigma_\mu{\cdot}\kappa = (1 - \theta^2/2i)\,\partial_\mu + i\theta^\nu\,\pi(L_{\mu\nu})$$

$$= (1 + \theta^2/2i)\,\partial_\mu + i\theta_\mu\theta{\cdot}\partial + i\theta^\nu\,\kappa_{\mu\nu}\,. \tag{3.3.9}$$

Here $\partial_\mu = \partial/\partial\theta^\mu$. Hence

$$s_\mu{}^\nu = (1 - i\theta^2/2)\,\delta_\mu{}^\nu + i\theta_\mu\theta^\nu\,,$$

$$\partial_\nu s_\mu{}^\nu = -2in\theta_\mu\,, \qquad s_\mu{}^\nu\,\theta_\nu = (1 + i\,\theta^2/2)\,\theta_\mu\,.$$

Let us calculate the invariant measure. First of all $M_j = 0$ in this case, and $q(x) = 1$. Eq. (2.2.24), $s_{\mu\nu}\,\omega_1[\xi] = 0$, means that ω_1 is a polynomial in θ^2, and Eq. (2.2.5) gives

$$\omega_1 = (1 + i\theta^2/2)^{2n}\,. \tag{3.3.10}$$

These formulas were first obtained long ago, by entirely different methods.[11] The case n = 2 was done earlier, by Keck.[9]

Note that the simple expressions for the operators $\pi(K_\mu)$ in (3.3.9) comes from having replaced ξ^μ by θ^μ according to (3.3.5). To take advantage of this, we express the superfield in terms of θ^μ, so that the transformation (1.3.10) takes the form

$$\Phi[K] = \int d_\ell\xi\; e^{\xi\cdot K} \otimes \Phi'[\theta]\,,$$

and the inner product (2.3.10) becomes

$$(\Phi_1, \Phi_2) = \int d_\ell\theta <\Phi_1'[\theta], \Phi_2'[\theta]> ,$$

$$d_\ell\theta = d\theta\,(1 - \theta^2/2i)^{2n} , \quad d\theta = d\theta^{2n} \ldots d\theta^1 .$$

3.4. Unitarizable representations.

The even part of osp(2n/1) is sp(2n,R), and the compact subalgebra is u(n). The center of u(n) is one-dimensional and generated by H/i, with

$$H = \frac{i}{4}\sum_\mu L_{\mu\mu} = \frac{1}{4}\sum_\mu K_\mu K_\mu . \tag{3.4.1}$$

In any representation π of osp(2n/1) in which the operators $\pi(K_\mu)$ are essentially selfadjoint, the operator $\pi(H)$ is positive.

Choose a compact Cartan subalgebra for sp(2n,R). A weight is a pair (E,w), where E is the projection on H and w is an su(n) weight. Fix a positive root system for su(n). We call (E,w) positive if E > 0 or if E = 0 and w < 0. Any irreducible, unitarizable representation π_0 of sp(2n,R) in which $\pi_0(H)$ is positive has a minimal weight (E_0,w_0), with multiplicity one and w_0 dominant integral, that characterizes it up to equivalence. We denote this representation $D(E_0,w_0)$.

<u>Theorem 3.4.2.</u> Every irreducible, unitarizable representation π of osp(2n/1) has a minimal weight (E_0,w_0) that characterizes it completely. The su(n) weight w_0 is dominant integral and the (E_0,w_0) weight space is one-dimensional.

Proof. Unitarizability implies that $\pi(K_\mu)$ is essentially selfadjoint and hence that $\pi(H)$ is positive. The center of sp(2n,R) acts on the odd part of osp(2n/1) by sending K_μ to $\pm K_\mu$; from this it follows that $\pi(H)$ has a purely discrete spectrum, and thus there is a smallest eigenvalue E_o of $\pi(H)$. The proof that the E_o eigenspace is an irreducible su(n) module goes precisely as for Lie algebras. Integrability on the compact subalgebra implies that this module is finite dimensional and hence determined by a dominant integral weight w_o. The fact that the representation is determined up to equivalence by (E_o,w_o) is proved as for Lie algebras.

We can thus label the irreducible, unitarizable representations of osp(2n/1) as $D_s(E_o,w_o)$--the suffix s (for super) distinguishing it from the corresponding representation of sp(2n,R). From now on we let $D(E_o,w_o)$ and $D_s(E_o,w_o)$ stand for the irreducible representations of sp(2n,R) and osp(2n/1), uniquely determined up to equivalence by a minimal weight (E_o,w_o), with w_o dominant integral, whether unitarizable or not. The problem is to determine, for each w_o, the range of values for E_o for which these representations are unitarizable. As far as sp(2n,R) is concerned, the complete answer is known.

It is a direct consequence of Corollary 1.2.8 that the reduction of $D_s(E_o,w_o)$ on $\mathfrak{g}_o$ = sp(2n,R) takes the form of a finite direct sum,

$$D_s(E_o,w_o)|sp(2n,R) = D(E_o,w_o) \oplus \sum_{k>0} \sum_{w} D(E_o + k/2, w). \quad (3.4.3)$$

It will be seen later that k ranges over a subset of the integers 1,...,n. Unitarizability of $D_s(E_o,w_o)$ clearly implies unitarizability of all the terms in this sum, and of $D(E_o,w_o)$ in particular. The converse statement is false.

A complete list of unitarizable representations of osp(4/1) has been given by Heidenreich.[10]

We investigate unitarizability within the family $\{D_s(E_0,w_0), E_0 \in R\}$ with w_0 fixed. We suppose that each component in (3.4.3) is unitarizable; this is certainly true if E_0 is large enough. We know that $D_s(E_0,w_0)$ has an invariant inner product, but we do not know whether it is positive. To verify positivity it is enough to select one vector from each $\mathfrak{g}_0$-module in (3.4.3), and check that these all have positive norms. If $|0>$ is a non-zero vector belonging to the minimal weight space (E_0,w_0), then such a set of vectors is

$$\{\pi(K_{\mu_1 \dots \mu_k})|0>\} , \quad k = 1,\dots,2n .$$

However, we can find a smaller set. Let

$$a_\mu = K_\mu + iK_{\mu+m} , \quad a_\mu{}^* = K_\mu - iK_{\mu+m} , \quad \mu = 1,\dots,n .$$

Then

$$[H,a_\mu] = -\frac{1}{2} a_\mu , \quad [H,a_\mu{}^*] = \frac{1}{2} a_\mu{}^* ,$$
$$[a_\mu,a_\nu{}^*] = (8/n)\, \delta_{\mu\nu} H + Q_{\mu\nu} . \tag{3.4.4}$$

Here $\{Q_{\mu\nu}\}$ $\mu,\nu = 1,\dots,n$ ($\Sigma\, Q_{\mu\mu} = 0$) is a basis for su(n).

Since the weight of $\pi(a_\mu)|0>$ is lower than that of $|0>$ we have

$$\pi(a_\mu)|0> = 0 , \quad \mu = 1,\dots,n .$$

A set of vectors that includes at least one from each $\mathfrak{g}_0$-irreducible component is given by

$$|\mu_1 ... \mu_k> = \pi(a_{\mu_1}{}^* ... a_{\mu_k}{}^*)|0> , \quad 1 \leq \mu_1 < \mu_2 ... < \mu_k \leq n ,$$

with $k = 0,1,...,n$. The eigenvalues of $\pi(H)$ on these vectors are $E_o + k/2$, which justifies Eq. (3.4.3). The invariant norms are

$$<0|\pi(a_{\mu_k} ... a_{\mu_1} a_{\mu_1}{}^* ... a_{\mu_k}{}^*)|0>$$

$$= [(8E_o/n)^k + ...] <0|0> . \qquad (3.4.5)$$

The bracket is a polynomial of order k in E_o, with coefficients determined by w_o and the leading term shown. Complete determination of these polynomials will give the full list of unitarizable representations of osp(2n/1). Meanwhile, the truth of the following is now obvious.

Theorem 3.3.6. Consider the family $\{D_s(E_o,w_o), E_o \in R\}$ of irreducible representations of osp(2n/1), with w_o fixed. There exists $|w_o|_s > 0$ such that $D_s(E_o,w_o)$ is unitarizable when $E_o \geq |w_o|_s$.

The best lower limit $E_o = |w_o|_s$ will be a zero of one of the polynomials in (3.4.5). It is the first reduction point of the "K-finite Verma module" and can be calculated from results obtained by Kac.[2] For additional information about unitarizable representations of osp(2n/1), see Ref. 15.

Acknowledgements

It is a pleasure to thank H. Araki, B. Binegar, R. Blattner, M. Flato, I. Ojima, and D. Sternheimer for helpful discussions. C. F. thanks H. Araki for hospitality at the Research Institute for Mathematical Sciences, and the Japanese Ministry of Education and the U.S. National Science Foundation for financial support.

References.

1. V. G. Kac, "Lie Superalgebras," Adv. in Math. 26, 8 (1977). See also P. G. O. Freund and I. Kaplansky, J. Math. Phys. 17, 228 (1976).
2. V. G. Kac, "Representations of Classical Lie Superalgebras," in "Differential Methods in Mathematical Physics" (K. Bleuler, H. R. Petry A. Reetzz, eds.), Springer-Verlag, Berlin 1978.
3. F. A. Berezin, The Method of Second Quantization (Acdemic Press, New York 1966) and Theor. Math. Phys. 6, 194 (1971).
4. F. A. Berezin and G. I. Kats, Mat. Sbornik 82, 124 (1970), English translation in 11, 311 (1970).
5. I. Bars, contribution to "School on Supersymmetry in Physics," Mexico 1981. A. B. Balantekin and I. Bars, J. Math. Phys. 22, 1149 and 1810 (1981).
7. A. Salam and J. Strathdee, Nucl. Phys. 76B, 477 (1974). See also J. Schwinger, Phys. Rev. 92, 1283 (1953).
8. W. Nahm, Nucl. Phys. B135, 149 (1978); Y. Ne'eman and S. Sternberg, Proc. Nat. Acad. Sci. 77, 3127 (1980); M. Marcu, J. Math. Phys. 11, 1277 (1980); P. H. Dondi and P. D. Jarvis, Zeit. für Physik C4, 201 (1980); A. B. Balantekin, I. Bars, and F. Iachello, Nucl. Phys. A370, 284 (1981).

9. Yu. A. Gol'fand and E. P. Likhtman, JETP Lett. 13, 323 (1971); A. Sala and J. Strathdee, Nucl. Phys. 80B, 499 (1974); B. W. Keck, J. Phys. A8 1819 (1975); B. L. Aneva, S. G. Mikhov, and D. Ts. Stoyanov, Teor. Math. Phys. 27, 502 (1976) and 31, 177 (1977) and 35, 383 (1978); B. Zumino, Nucl. Phys. B127, 189 (1977); E. A. Ivanov and A. S. Sorin Teor. Math. Phys. 45, 862 (1980) and J. Phys. 13, 1159 (1980); E. Sokatchev, Nucl. Phys. B99, 96 (1975); P. D. Jarvis, J. Math. Phys. 17, 916 (1976).

10. W. Heidenreich, Phys. Lett. 110B, 461 (1982); M. Günaydin and C. Saclioglu, Phys. Lett. 108B, 169 (1982); P. Breitenlohner and D. Z. Freedman, Ann. Phys. 144, 249 (1982); M. Flato and C. Fronsdal, Lett. Math. Phys. 8, 159 (1984); D. Z. Freedman and H. Nicolai, Nucl. Phys. B237, 342 (1984); L. Castell, W. Heidenreich and T. Künemund, "All Unitary Positive Energy UIR's of osp(N,4)," Starnberg preprint 1984; M. Günaydin, P. van Nieuwenhuizen and N. P. Warner, Nucl. Phy B255, 63 (1985).

1. C. Fronsdal, Lett. Math. Phys. 1, 165 (1976); Phys. Rev. D26, 1988 (1982).

2. R. Arnowitt and P. Nath, Phys. Lett. 56B, 177 (1975); R. Arnowitt, P. Nath and B. Zumino, ibid. p. 81.

3. C. Fronsdal, "3+2 de Sitter Superfields," UCLA preprint May 1985, following paper.

4. C. Fronsdal, "Semisimple Gauge Theories and Conformal Gravity," Lectures in Applied Mathematics 21, 165 (1985); H. Araki, Comm. Math Phys. 97, 149 (1985).

5. B. S. De Witt, "Supermanifolds," Cambridge University Press, New York 1984.

3+2 DE SITTER SUPERFIELDS

by

C. Fronsdal

ABSTRACT. An efficient treatment of de Sitter supersymmetry makes it possible to present a rather complete group theoretical analysis of de Sitter superfields. Much of the analysis is independent of the type and number of indices on the superfield. The covariant derivatives form a 5-dimensional representation of osp(4/1). The invariant operator formed from the covariant derivatives (the Dirac operator) provides, by its spectral decomposition, a reduction of the field into irreducible parts (in the massive case). Greatest emphasis is placed on the massless case, and a complete account is presented of the interesting nondecomposable representations that appear. The vector multiplet is described by a superfield analog of the Fierz-Pauli equation. Its massless limit (super QED) is analyzed in detail, with its full gauge structure and indefinite metric quantization. A chirality operator is also described, and a preliminary discussion of N = 2 extended superfields. A strong analogy between N = 2 supersymmetry and conformal invariance is pointed out and related to the difficulties that still confront the completion of N = 2 superspace field theory.

C. Fronsdal (ed.), Essays on Supersymmetry, 67–122.

0. Introduction

Physics in de Sitter space has not yet been developed to the point that one can describe, say, electroweak theory or phenomenology; this in spite of the fact that the recalcitrant cosmological constant is commanding much attention. Even less is widely known, as can be seen from the treatment accorded the cosmological constant in the relatively recent literature. One reason for the lack of propagation of knowledge is the curious resistance to group theoretical methods and concepts. Thus it seems to have come as a surprise, for example, that stability of some field theories[1] is equivalent to the unitarity of a representation of the space time group. Another example: Hermiticity of a group theoretical generator, the Hamiltonian, is not always readily understood to be an expression of energy conservation.[2] This reluctance to make use of a tremendously effective tool is evident in the context of flat space physics also. Although group theoretical methods are standard in the treatment of massive superfields, they are relatively untried in the more interesting massless case. [About this Taylor says: "This proves difficult to discuss directly due to the extra gauge invariance."[3]] See, however, the work of Okubo.[4]

One purpose of this paper is to demonstrate the power of group theoretical methods to handle the important and interesting phenomena that are associated with massless fields. But why in de Sitter space? It is by now fairly obvious that de Sitter space is "relevant," as seen for example in recent developments in supergravity[4] and Kaluza-Klein theories.[5] But there is a more important reason: massless fields are even more interesting in de Sitter space than in flat space. The theory of unitary representations of the 3+2 de Sitter group exhibits a fine structure in the infrared region that is without parallel in the case of the

Poincaré group. It is enough to mention the exciting singletons[6] and the curious duplicity of QED.[7,8]

One reason for the relative lack of development of de Sitter supersymmetry has been the unavailability of an efficient notation. We hope to supply that here. The analysis of de Sitter superfields presented here is essentially independent of the number and type of indices that they carry, and much of the general analysis is only slightly more complicated than in the case of the scalar (index-free) superfield. One of the essential points is the avoidance of coordinates; it is much simpler to work directly with the structure. Another important point is to rely entirely on the symplectic notation, sp(4/R) instead of so(3,2). This avoids the complications of Fierzing and the charge conjugation matrix. It also simplifies the passage to the flat space limit.

This paper, intended as a beginning, deals in full detail with the scalar superfield in N = 1 supersymmetry. The extension to N = 2 is in some respects very simple, and those aspects are dealt with, but in other respects they are very complicated. We hope to return to the more difficult aspects in the future. The concept of de Sitter chirality, very important in the context of super Yang-Mills and supergravity, is developed in the last sections.

1. Superfields and Induced Representations

The starting point is a formal representation of sp(4,R) defined by operators $\kappa_{ab} = \kappa_{ba}$, a,b = 1,...,4, and the mapping

$$L_{ab} \rightarrow \kappa_{ab} \, .$$

The principal example that we have in mind is the realization of sp(4,R) by vector fields on 3+2 de Sitter space; more precisely the natural action that arises from the isomorphism between sp(4,R) and so(3,2). However, it is not convenient to specialize too early; de Sitter space can be replaced by any sp(4,R)-homogeneous space M, and (κ_{ab}) can be taken to stand for the action of sp(4,R) on any vector bundle over M. Scalar, tensor and spinor superfields are thus treated together and all at once. Later we shall restrict this representation to a subspace V_o and turn the κ_{ab} into the operators $\pi_o(L_{ab})$ of an irreducible representation π_o of sp(4,R) in V_o. For the present, V_o will indicate any space of sections of some vector bundle over M.

The symmetric algebra S_1 over the odd part of the superalgebra osp(4/1) is a Grassmann algebra generated by $\{\theta^a\}$ a = 1,...,4 and a unit. [To be quite precise, if S_1 is the symmetric algebra of the odd part of osp(4/1), then the algebra generated by the θ's is the dual $S_1{}^*$. Our notation does not distinguish between S_1 and $S_1{}^*$.] The adjoint action of sp(4,R) on the odd part of osp(4/1) gives rise to the action

$$L_{ab} \to s_{ab} = - \theta_a \partial_b - \theta_b \partial_a \tag{1.1}$$

on S_1. A V_o superfield Φ is a polynomial in $\{\theta^a\}$ with coefficients in V_o; in other words, a section of the vector bundle $S_1 \otimes V_o$ over M. The natural interpretation of this object in terms of induced representations leads to the following action of osp(4/1) on Φ:

$$L_{ab} \to \pi(L_{ab}) = s_{ab} + \kappa_{ab} , \tag{1.2}$$

$$K_a \to \pi(K_a) = (1 + i\theta^2/2)\, \partial_a + i\theta^b\, \pi(L_{ab}) . \tag{1.3}$$

These formulas were actually obtained long ago by an entirely different construction, and only recently related to induced representations. A simple derivation may be found in the Appendix. We shall work with a slightly modified form of (1.3),

$$\pi(K_a) = (1 + i\theta^2/2)\,\partial_a + i\theta^b\,\pi(L_{ab}) - 2i\theta_a\,. \qquad (1.4)$$

The last term (a multiplier) is introduced in order to simplify the invariant Berezin measure. It can be removed by a simple equivalence transformation; see Appendix, Section A3.

The use of induction to construct unitarizable representations of superalgebras goes back to Gol'fand and Likhtman,[9] and was used to very good effect by Salam and Strathdee.[10] Dondi and Sohnius[11] applied the method to su(2,2/2), and Keck[12] used it for osp(4/1). A very recent paper investigates induced representations of supergroups in general and osp(2n/1) in particular.[13]

2. Induction From An Irreducible Representation

Here, let V_o be a space of sections determined by field equations and boundary conditions, the space of free field modes. We suppose that the energy is bounded below, so that the representation π_o of so(4,R) in V_o, defined by $L_{ab} \to \pi_o(L_{ab}) = \kappa_{ab}$, has a minimal weight. We also suppose that this representation is irreducible; hence

$$\pi_o = D(E_o,s)\,,$$

where (E_o,s) is the minimal weight. The notation is standard; E_o denotes the minimal energy, and the spin s is the angular momentum of the ground state.

We now calculate the induced representation

$$\pi = \mathrm{IND} \underset{sp(4,R)}{\overset{osp(4/1)}{\uparrow}} \pi_o . \tag{2.1}$$

The restriction to sp(4,R) is

$$\pi|sp(4,R) = S_1 \otimes \pi_o , \tag{2.2}$$

where S_1 is the Grassmann algebra defined in Section 1, here considered as an sp(4,R) module. If $\underline{4}$ denotes the 4-dimensional, defining representation of sp(4,R), then

$$S_1 = \mathrm{Id} \oplus \underline{4} \oplus (\underline{4} \wedge \underline{4}) \oplus \underline{4} \oplus \mathrm{Id} . \tag{2.3}$$

The reduction of the direct product is easy to calculate when $E_o > s + 2$ (massive case), for then the minimal weight (E_o,s) is safely inside the positive Weyl chamber. Then the irreducible components of (2.2) are

$$\mathrm{Id} \otimes D(E_o,s) = D(E_o,s) \qquad , \text{ (twice)}$$

$$\underline{4} \otimes D(E_o,s) = D(E_o + \tfrac{1}{2}, s + \tfrac{1}{2}) \oplus D(E_o + \tfrac{1}{2}, s - \tfrac{1}{2})$$

$$\oplus D(E_o - \tfrac{1}{2}, s + \tfrac{1}{2}) \oplus D(E_o - \tfrac{1}{2}, s - \tfrac{1}{2}) , \text{ (twice)}$$

$$(\underline{4} \wedge \underline{4}) \otimes D(E_o,s) = D(E_o+1,s) \oplus D(E_o,s+1) \oplus D(E_o,s)$$

$$\oplus D(E_o,s-1) \oplus D(E_o-1,s) \oplus D(E_o,s) . \tag{2.4}$$

The terms $D(E_o',s')$ for which $s' < 0$ are to be ignored. The last term in the last formula should be dropped when s = 0.

These formulas remain valid for lower values of E_o, except at the reduction points. The exceptional values of E_o are associated with massless representations[7,14] and singletons,[6] and we must therefore take a special interest in them. If s = 0 and $E_o \geq 0$, then (2.4) holds except in the following cases.

$$\underline{4} \otimes D(\tfrac{3}{2},0) = D(2,\tfrac{1}{2}) \to D(1,\tfrac{1}{2}) \to D(2,\tfrac{1}{2}) ,$$

$$(\underline{4} \wedge \underline{4}) \otimes D(\tfrac{3}{2},0) = D(\tfrac{3}{2},1) \oplus D(\tfrac{3}{2},0)$$

$$\oplus\ [D(\tfrac{5}{2},0) \to D(\tfrac{1}{2},0) \to D(\tfrac{5}{2},0)] ,$$

$$(\underline{4} \wedge \underline{4}) \otimes D(2,0) = D(2,0) \oplus D(1,0)$$

$$\oplus\ [D(3,0) \to D(2,1) \to D(3,0)]$$

$$(\underline{4} \wedge \underline{4}) \otimes D(1,0) = D(2,0) \oplus D(1,0)$$

$$\oplus\ [D(1,1) \to \{D(2,1) \oplus Id\} \to D(1,1)] . \qquad (2.5)$$

The complete reduction of (2.2) is shown in Fig. 1a (generic case), Fig. 1b (singletons) and Figs. 1c, 1d (electrodynamics). The physical, transverse photon module is D(2,1), and the singletons are $D(\tfrac{1}{2},0)$ and $D(1,\tfrac{1}{2})$. The components of the non-decomposable factors are connected by broken lines. The dotted lines indicate the Weyl planes through the point ρ = the half-sum of the negative roots of sp(4,R).

The irreducible, minimal weight representations of osp(4/1) will be denoted $D^s(E_o,s)$. To determine the structure of the induced representation (2.1) of osp(4/1) we need to know the reduction of

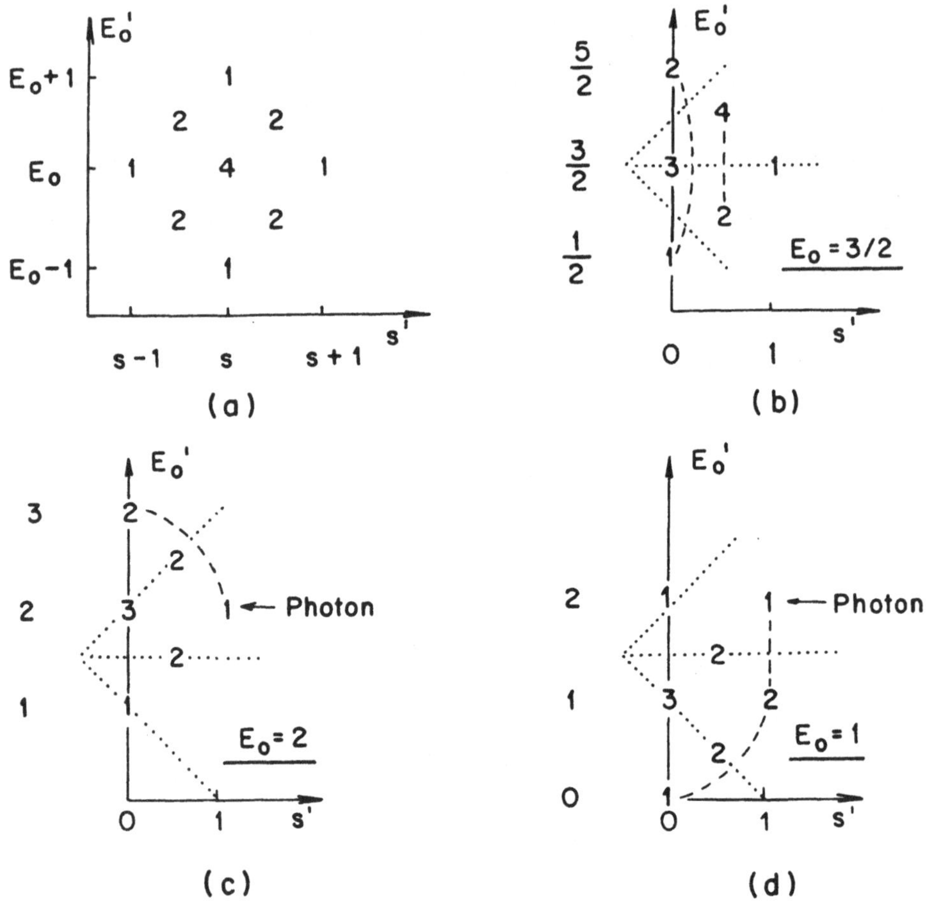

Fig. 1.

The multiplicities of the reduction of $S_1 \otimes D(E_0,s)$ as a sum over $D(E_0',s')$'s is shown in (a) for the generic case. The points with negative values of s' should be ignored, and in the special case when s = 0 the multiplicity of (E_0,s) in (a) is 3 instead of 4. In the exceptional cases (b), (c) and (d) the number of sp(4,R)-irreducible subquotients is shown. Broken lines connect the components of non-decomposable factors. Dotted lines indicate the Weyl planes. The representations encountered in electrodynamics are shown in (c) and (d), with the physical, transverse photon representation emphasized.

$D^s(E_0,s)$ on sp(4,R). In the generic case[15]

$$D^s(E_0,s)|_{sp(4,R)} = D(E_0,s) \oplus D(E_0 + \tfrac{1}{2}, s + \tfrac{1}{2})$$

$$\oplus\, D(E_0 + \tfrac{1}{2}, s - \tfrac{1}{2}) \oplus D(E_0 + 1,s)\,. \quad (2.6)$$

The third term is to be ignored when s = 0. The exceptional cases of interest are as follows: for $D^s(\frac{1}{2},\frac{1}{2})$ the last term is absent, for $D^s(\frac{1}{2},0)$ and for $D^s(\frac{3}{2},\frac{1}{2})$ the last two terms are absent, for $D^s(0,0)$ only the first term remains.

Now we have enough information to determine the subquotients of the induced representation (2.1). In the generic case we have full reducibility, and simply by balancing the multiplicities of the sp(4,R) submodules we find that

$$\mathrm{IND}\underset{sp(4,R)}{\overset{osp(4/1)}{\uparrow}} D(E_0,s) = D^s(E_0 - 1, s) \oplus D^s(E_0 - \tfrac{1}{2}, s - \tfrac{1}{2})$$

$$\oplus\, D^s(E_0 - \tfrac{1}{2}, s + \tfrac{1}{2}) \oplus D(E_0,s)\,. (2.7)$$

As always, terms with negative spin are to be ignored.

The exceptional cases (for s = 0) are $E_0 = \frac{3}{2}$, 2 and 1. The subquotients can easily be found in the same way, but some of the minimal weights become super Weyl equivalent, and non-decomposable subrepresentations appear.

Thus, for $E_0 = 3/2$, the subquotients are $D^s(\frac{3}{2},0)$ (twice), $D^s(\frac{1}{2},0)$ and $D^s(1,\frac{1}{2})$. The first three are super Weyl equivalent; that is, they have the same infinitesimal character. Restriction to sp(4,R) shows that the induced representation is not fully reducible. In fact, the information in (2.5) is enough to establish the following result:

$$\mathrm{IND}\underset{sp(4,R)}{\overset{osp(4/1)}{\uparrow}} D(\tfrac{3}{2},0) = [D^S(\tfrac{3}{2},0) \to D^S(\tfrac{1}{2},0) \to D^S(\tfrac{3}{2},0)]$$

$$\oplus D^S(1,\tfrac{1}{2}) , \tag{2.8}$$

$$\mathrm{IND}\uparrow D(2,0) = [D^S(2,0) \to D^S(\tfrac{3}{2},\tfrac{1}{2}) \to D^S(2,0)]$$

$$\oplus D^S(1,0) , \tag{2.9}$$

$$\mathrm{IND}\uparrow D(1,0) = [D^S(\tfrac{1}{2},\tfrac{1}{2}) \to \{D^S(\tfrac{3}{2},\tfrac{1}{2}) \oplus \mathrm{Id}\} \to D^S(\tfrac{1}{2},\tfrac{1}{2})]$$

$$\oplus D^S(1,0) . \tag{2.10}$$

Each representation contains a complete Gupta-Bleuler triplet. Since all representations that are induced from the even part of the superalgebra admit a nondegenerate, invariant Hermitean form,[13] this could have been predicted on the basis of Araki's theorem.[16] [This theorem was proved for Lie algebras, but it is applicable to superalgebras as well.] Fig. 2 illustrates the result, and Fig. 3 attempts to combine the information in Figs. 1 and 2.

The reduction (2.6) was obtained by Heidenreich.[15] The induced representations (2.9) and (2.10) are the zero center modules of Ref. 17. The representations $D^S(1,0)$, $D^S(2,0)$ and $D^S(\tfrac{3}{2},\tfrac{1}{2})$ were identified by Breitenlohner and Freedman[18] in their study of de Sitter supergravity.

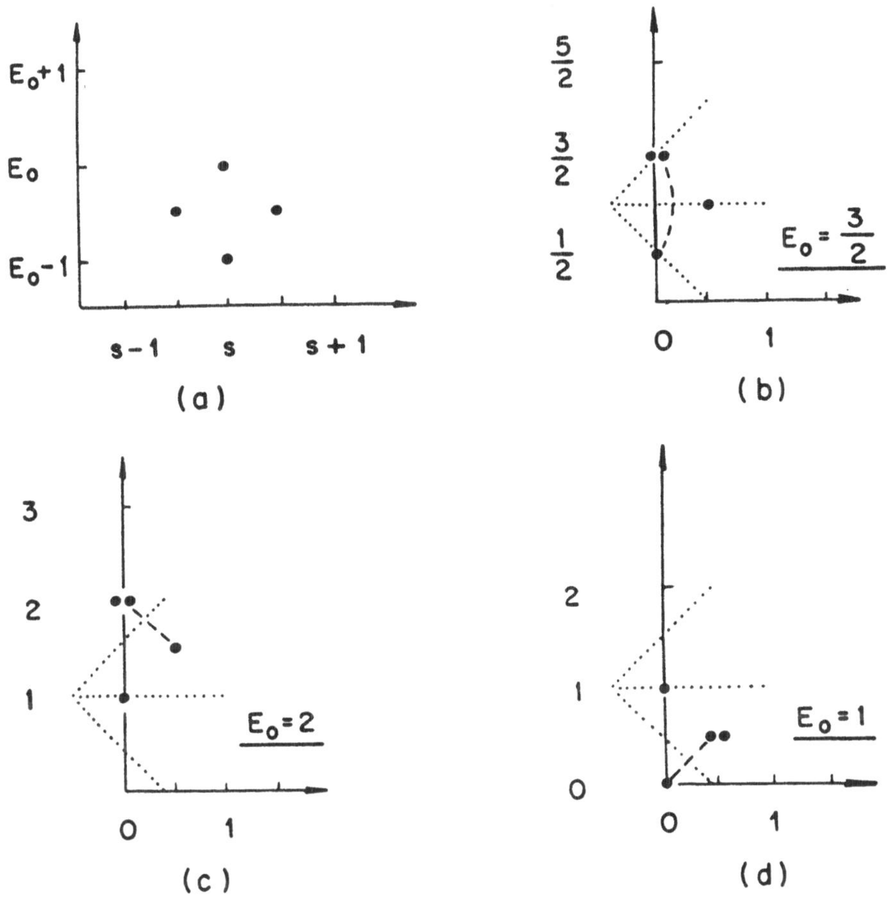

Fig. 2.

The reduction of the induced representation. Each dot represents an irreducible subquotient. The inducing representation is $D(E_0,s)$ in the generic case, Fig. 1(a). Here the dot at the left should be suppressed if $s = 0$. Broken lines connect the subquotients of the super Gupta-Bleuler triplets, and dotted lines indicate the super Weyl planes.

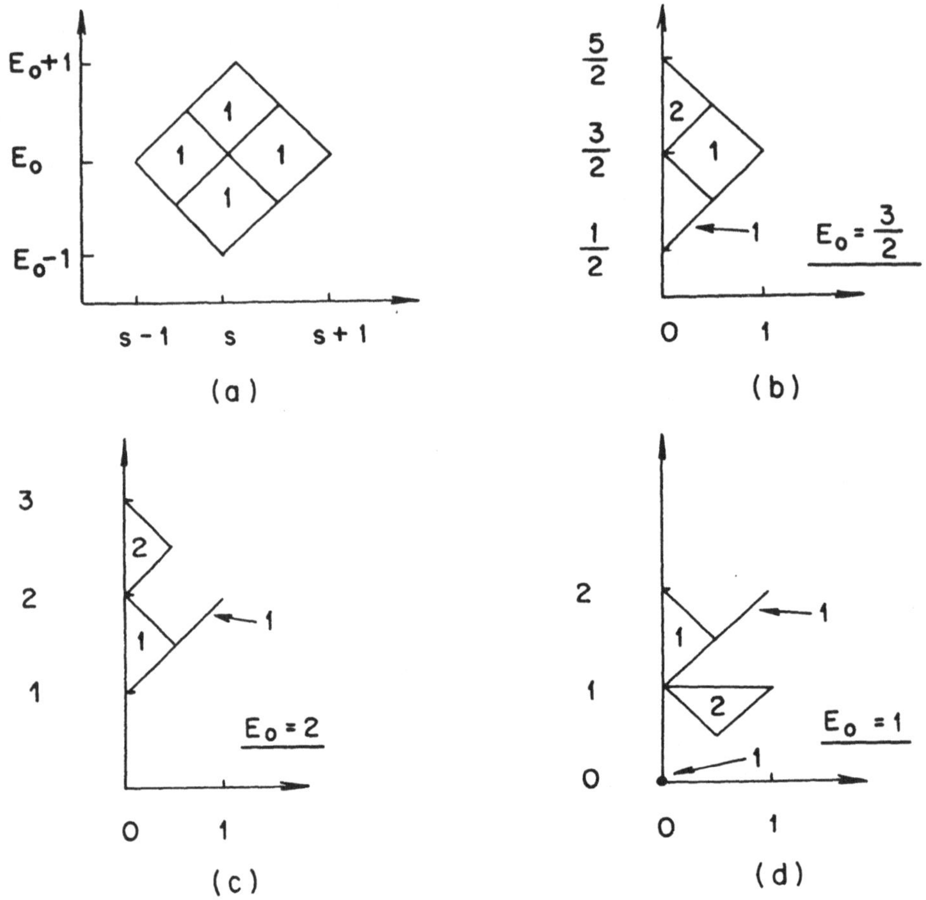

Fig. 3.

This combines the information in Figs. 1 and 2. Each "unit" stands for an irreducible representation of osp(4/1), and each is marked with the multiplicity with which it occurs in the induction from the representation $D(E_0,s)$ of sp(4,R). In (b), (c), (d) the inducing spin is zero. The "units" are: diamonds, triangles, line segments and in one instance a point. The corners or end points of each unit represent the reduction of each irreducible osp(4/1) submodule on sp(4,R). These diagrams are the osp(4/1) counterparts of O'Raifeartaigh's super Poincaré weight diagrams.[19]

3. Invariant Operators

Eq. (2.6) shows that any particular irreducible representation of osp(4/1) can be realized by projection from a superfield in 4 different ways. At least this is true in the generic (massive) case; the exceptional (massless) cases must be treated separately. We shall now construct the projection operators explicitly. More precisely, we shall find a second order differential operator, the spectral resolution of which coincides with the reduction (2.6) of the superfield into irreducible parts. This operator will give us the field equations for the superfield.

Two invariant second order differential operators are available. First, the Casimir operator of the inducing representation π_o of sp(4,R),

$$\Box = \frac{1}{4} \kappa_a{}^b \kappa_b{}^a , \tag{3.1}$$

and secondly the Casimir operator of the induced representation π of osp(4/1)

$$\mathcal{C}_1 = \frac{1}{4} \pi(L_a{}^b L_b{}^a - i K^a K_a) . \tag{3.2}$$

The operator that we are looking for will turn out to be a linear combination of these two.

To evaluate (3.2) on Φ we begin by expanding the superfield,

$$\Phi = \phi + \theta{\cdot}\psi + i\theta^2 A + \frac{i}{2} \theta^{ab} A_{ab} + i\theta^2 \theta{\cdot}\chi + \theta^4 F , \tag{3.3}$$

with ϕ, ψ_a, A, A_{ab}, χ_a, F in V_o and

$$\theta^{ab} = \theta^a \theta^b + \eta^{ab} \theta^2/4 , \quad \eta^{ab} A_{ab} = 0 .$$

The action of (3.1) and (3.2) on the components of Φ is by second order differential operators. In the simplest case ϕ, A and F are scalar fields, while ψ and χ are spinor fields. This suggests that the wave operator ought to act as a first order differential operator on ψ and χ. There is no invariant operator with that property, so we have to ask for less.

Definition. A Dirac operator is an invariant operator that acts as a first order differential operator on the components ψ_a of Φ.

It turns out that such an operator exists, and that it is essentially unique. It is a linear combination of the two Casimir operators (3.1) and (3.2). Since the direct evaluation of (3.2) would be lengthy, we shall arrive at our Dirac operator by a different route.

In Poincaré supersymmetry an important operator is the spinorial "covariant derivative."[20] In the case of osp(4/1) this operator has the following close analogue. Let $\{Q_a\}$ $\alpha = 0,1,\dots,4$ be the 5 operators defined by

$$Q_o = 4 - 2\theta{\cdot}\partial \; , \tag{3.4}$$

$$Q_a = i(K_a - 2\partial_a) \; , \quad a = 1,\dots,4 \; . \tag{3.5}$$

From now on we write K_a and L_{ab} instead of the more cumbersome $\pi(K_a)$ and $\pi(L_{ab})$ for the action of the generators on the superfield. One easily verifies that

$$[L_{ab},Q_o] = 0 \; , \quad [L_{ab},Q_c] = -\eta_{bc}Q_a - \eta_{ac}Q_b \; ,$$

$$[K_a,Q_o] = -2iQ_a \; , \quad [K_a,Q_b]_+ = \eta_{ab}Q_o \; . \tag{3.6}$$

This determines an action of the superalgebra on the five Q_α's, and hence an irreducible, 5-dimensional representation of osp(4/1).

The operator

$$\mathcal{D} \equiv \frac{1}{2i}(Q^a Q_a + \frac{1}{2i} Q_o^2) \tag{3.7}$$

commutes with K_a and with L_{ab}. Later we shall verify that

$$\mathcal{D} = 4 + 2\Box - 2\mathcal{C}_1 . \tag{3.8}$$

Direct evaluation on (3.3) gives the explicit form of the action of K_a, Q_a and $\mathcal{D}$ on the superfield,

$$\left.\begin{array}{l} K_a \Phi \\ -i Q_a \Phi \end{array}\right\} = \pm\psi + i\theta^b [\kappa_{ab}\phi + 2\eta_{ab}\phi \mp 2\eta_{ab}A \pm A_{ab}]$$
$$+ (i/4)\, \theta^2 [(\kappa-1)\psi \pm 2\chi]_a$$
$$+ i\theta^{bc} [\kappa_{ab}\psi_c + \eta_{ab}\psi_c \mp 2\eta_{ab}\chi_c]$$
$$+ \theta^2\theta^b \left[\frac{1}{2}(\kappa-1)_a{}^d A_{bd} - \kappa_{ab}A - \eta_{ab}A \mp 4\eta_{ab}F\right]$$
$$+ (1/4)\, \theta^4 (2-\kappa)_a{}^b \chi_b , \tag{3.9}$$

$$\mathcal{D}\Phi = 4A + \theta^a [(\kappa-1)\psi + 2\chi]_a + i\theta^2 [(1+\Box/2)\phi + 2A - 4F]$$
$$+ i\theta^{ab} [(\kappa-1)_a{}^c A_{cb} - \frac{1}{2}(\kappa\kappa)_{ab}\phi]$$
$$+ i\theta^2\theta^b [\Box\psi_b + (1/4)\{(\kappa\kappa)_{bd} - (\kappa\kappa)_{db}\}\psi^d$$
$$+ \frac{1}{2}(\kappa+1)_b{}^c \psi_c + (\kappa-1)_b{}^c \chi_c]$$
$$- \theta^4 [(1+\Box/2)A + (1/16)(\kappa\kappa)^{ab} A_{ab}] . \tag{3.10}$$

We see that $\mathcal{D}$ satisfies our definition of a Dirac operator. Henceforth it will be referred to as the Dirac operator.

It should be emphasized that these formulas apply to spinor and tensor superfields as well as to scalar superfields. There is only a minor simplification in the case of scalar superfields; that is, when the inducing spin s = 0. In this case

$$\kappa_a{}^b \kappa_b{}^c = \delta_a{}^c \Box + 3\kappa_a{}^c ,$$

and (3.10) reduces to

$$\begin{aligned}\mathcal{D}\Phi = 4A + \theta^a [(\kappa-1)\psi + 2\chi]_a + i\theta^2 [(1+\Box/2)\phi + 2A - 4F] \\ + i\theta^{ab} (\kappa-1)_a{}^c A_{cb} \\ + (i/2)\theta^2 \theta^a (\kappa-1)_a{}^b [(\kappa-1)\psi + 2\chi]_b \\ - \theta^4 (1+\Box/2) A . \qquad (3.10')\end{aligned}$$

The values of $\Box$ and $\mathcal{C}_1$ are:

$$\begin{aligned}&\text{In } D(E_o,s) : \Box = E_o(E_o-3) + s(s+1) , \\ &\text{In } D^s(E_o,s) : \mathcal{C}_1 = E_o(E_o-2) + s(s+1) . \end{aligned} \qquad (3.11)$$

Consider the case when the inducing representation π_o is irreducible,

$$\pi_o = D(E_o,s) .$$

Then from (2.7), (3.8) and (3.11) we deduce that the eigenvalues of $\mathcal{D}$ in the four irreducible components of the induced representation are as follows:

$$D^s(E_0-1,s)\,,\ D^s(E_0-\tfrac{1}{2},s-\tfrac{1}{2})\,,\ D^s(E_0-\tfrac{1}{2},s+\tfrac{1}{2})\,,\ D^s(E_0,s)\,,$$
$$2E_0-2\qquad,\ 2s+2\qquad\qquad,\ -2s\qquad\qquad\qquad,\ 4-2E_0\ . \tag{3.12}$$

In the generic case these 4 eigenvalues are distinct. The second term is absent when s = 0.

The eigenvalues give us the minimal polynomial, namely

$$P(\mathcal{D}) = \mathcal{D}\,[\mathcal{D}^2 - 2\mathcal{D} - 8 - 4E_0(E_0-3)]$$

when s = 0, and

$$P(\mathcal{D}) = \mathcal{D}\,(\mathcal{D}-2)\,[\mathcal{D}^2 - 2\mathcal{D} - 8 - 4E_0(E_0-3) - 4s(s+1)]$$

$$+\ 16s(s+1)E_0(E_0-3)$$

when $s \neq 0$. This gives us an operator identity. When s = 0,

$$\mathcal{D}\,[\mathcal{D}^2 - 2\mathcal{D} - 8 - 4\Box] = 0\ . \tag{3.13}$$

When $s \neq 0$, we need the 4th order Casimir operator of π_0, or more precisely the operator $\Box'$ that is defined by its value:

$$\text{In } D(E_0,s)\ :\ \Box' = s(s+1)(E_0-1)(E_0-2)\ . \tag{3.14}$$

The operator identity for the case that $s \neq 0$ takes the form

$$\mathcal{D}\,(\mathcal{D}-2)\,[\mathcal{D}^2 - 2\mathcal{D} - 8 - 4\Box] + 16\Box' = 0\ . \tag{3.15}$$

The operator algebra (3.6) was discovered in Ref. 21 where it had the interpretation as the conformal algebra of a superspace with constant curvature. The structure is that of osp(4/2). In the flat space limit the last term in (3.7) no longer contributes, and $\mathcal{D}$ reduces to the operator that is usually denoted $\bar{D}D$.[20] The factor $\mathcal{D}^2 - 2\mathcal{D} - 8 - 4\Box$ becomes $-4\Box$ times the vector multiplet projection operator.[22] These remarks apply to the case of the scalar superfield. The chiral decomposition $\bar{D}D = \bar{D}_+D_- + \bar{D}_-D_+$ is possible in de Sitter space also, see Section 9.

4. Massive Superfields, "Scalar" Multiplet

Since $\mathcal{D}$ is essentially self-adjoint when π_0 is unitarizable, we may consider the simple action

$$\int dx d\theta\, \Phi^\dagger(\mathcal{D} - 2m)\, \Phi \tag{4.1}$$

and the wave equation

$$(\mathcal{D} - 2m)\, \Phi = 0\,. \tag{4.2}$$

We suppose that the value of s has been fixed by the choice of the type of inducing representation. However, since it is our intention to describe a superfield that is not a priori on-shell, we do not fix π_0. Normally, π_0 will be highly reducible, and the wave equation will pick out one irreducible component.

Positive energy solutions of (4.2) may carry representations of the type $D^s(E_0',s')$, provided that 2m coincides with one of the eigenvalues of $\mathcal{D}$. The information provided in (3.12) tells us the following:

(i) If $m > s + 1$, then only the first entry in (3.12) gives us a unitarizable representation. We must take the inducing representation to be $D(m+1,s)$ and the physical modes will then carry $D^S(m,s)$.

(ii) If $m < -s$, then only the last entry in (3.12) is applicable. We must take the inducing representation to be $D(2-m,s)$ and the physical modes will then carry $D^S(2-m,s)$.

Referring to Fig. 3a, we have thus succeeded in isolating the lower-square irreducible part of the superfield by taking $m > s + 1$, and the upper-square irreducible part by taking $m < -s$. There are thus two different ways of realizing one and the same irreducible, massive representation. What about the remaining two irreducible parts of the superfield? Wave equations that isolate these will be discussed in the next section. We shall also deal with the exceptional cases $m = s + 1$ and $m = -s$. For the present we concentrate on the regime $m > s + 1$.

Although Eq. (4.2) is quite simple for any spin, we shall write it down in component form in the case $s = 0$ only. The scalar fields satisfy

$$[\Box + 2 - m(m-1)]\,\phi = 0 \quad ,$$

$$2A = m\phi \;, \quad -8F = m(m-1)\,\phi \;. \tag{4.3}$$

For the spinor fields we find

$$2\chi = m\psi \;, \quad (\kappa - 1 - m)\,\psi = 0 \;, \tag{4.4}$$

and the wave equation for the vector field is

$$(\kappa - m - 1)_a{}^c\, A_{cb} - (a,b) = 0 \;. \tag{4.5}$$

To interpret the last equation it helps to translate to a more familiar notation.

Let (y_α) $\alpha = 0,1,2,3,5$ be coordinates for R^5, and realize de Sitter space as the universal covering of the hyperboloid

$$y^2 = \delta^{\alpha\beta}\, y_\alpha y_\beta = y_0^{\,2} - y_1^{\,2} - y_2^{\,2} - y_3^{\,2} + y_5^{\,2} = 1 .$$

Let (γ_α) $\alpha = 0,1,2,3,5$ be 4-dimensional, real Dirac matrices,

$$[\gamma_\alpha, \gamma_\beta]_+ = -2\delta_{\alpha\beta} .$$

In this notation,

$$\Box = -y^2\, \partial^2 + y{\cdot}\partial(y{\cdot}\partial + 3) ,$$

$$(\kappa_a{}^b) = -\gamma{\cdot}y\; \gamma{\cdot}\partial - y{\cdot}\partial .$$

Now let

$$A\!\!\!/ = \gamma{\cdot}A = (A_a{}^b)$$

define the vector field (A_α). Equivalently

$$\gamma_\alpha{}^{ab}\, A_{ab} = \mathrm{tr}\, \gamma_\alpha A\!\!\!/ = -4A_\alpha .$$

The wave equation (4.7) now takes the form

$$\partial_\alpha y{\cdot}A - y_\alpha \partial{\cdot}A + mA_\alpha = 0 ,$$

from which it is seen that our vector field is determined by two scalar fields and that these in turn satisfy

$$y^2\,\partial\cdot A = (y\cdot\partial + m)\,y\cdot A\ ,$$

$$[\square - m(m-3)]\,y\cdot A = 0\ .$$

The dynamically independent fields ϕ, ψ, $y\cdot A$ carry the irreducible representations $D(m+1,0)$, $D(m+\frac{1}{2},\frac{1}{2})$, $D(m,0)$, respectively.

In the flat space limit Eq. (4.2) reads

$$(\bar{D}_+ D_- + \bar{D}_- D_+ - 2m)\,\Phi = 0\ .$$

The solution is a sum of a chiral and an antichiral field,

$$\Phi = \Phi_+ + \Phi_-$$

$$D_-\Phi_+ = 0 = D_+\Phi_-\ .$$

The wave equation now reduces to

$$\bar{D}_+ D_- \Phi_- = 2m\Phi_+\ ,\quad \bar{D}_- D_+ \Phi_+ = 2m\Phi_-\ ,$$

from which it easily follows that

$$(\partial^2 + m^2)\,\Phi_- = 0 = (\partial^2 + m^2)\,\Phi_+\ .$$

However, the interesting finer details of the approach to the flat space limit have not been analyzed. In view of the fact that the contraction must discriminate between positive and negative helicities[14] it is not surprising if the limit theory is unsymmetrical.

So far, we have not encountered spins other than 0 and 1/2. To describe spin-1 fields we have two options. The obvious way is to take the inducing representation π_0 to have spin; $s = \frac{1}{2}$ would do. The other way is more interesting and will be described in the next section.

5. The "Vector" Multiplet

Let us restrict ourselves to the case of a spinless inducing representation. The middle square in Fig. 3b is associated with the eigenvalue zero of $\mathcal{D}$. But $\mathcal{D}\Phi = 0$ is not a wave equation. Unlike Eq. (4.2) with $m \neq 0$, it does not fix the mass parameter E_0. It is just a subsidiary condition, and it must be supplemented by a true wave equation. In fact, however, we need a Fierz-Pauli equation that incorporates both. The Fierz- Pauli wave operator must be annihilated by the subsidiary condition, and the identity (3.13),

$$\mathcal{D}\mathcal{M} = 0\ , \quad \mathcal{M} \equiv \mathcal{D}^2 - 2\mathcal{D} - 8 - 4\Box \tag{5.1}$$

suggests the wave equation

$$(\mathcal{M} + 4m^2 - 1)\,\Phi = 0\ . \tag{5.2}$$

In view of (5.1) it implies that, if $4m^2 \neq 1$,

$$\mathcal{D}\Phi = 0 \tag{5.3}$$

$$(\Box + \frac{9}{4} - m^2)\,\Phi = 0\ . \tag{5.4}$$

The inducing spin was taken to be zero, and the value of $\Box$ in $D(E_0,0)$ is

$E_0(E_0 - 3)$. Therefore, (5.4) shows that the positive energy inducing representation is $D(E_0,0)$ with $E_0 = m + 3/2$. Since $4m^2 \neq 1$, $E_0 \neq 1$ and $E_0 \neq 2$. A glance at (3.12) shows that (5.3) holds only in the irreducible representation $D^S(E_0 - \frac{1}{2}, \frac{1}{2}) = D^S(m + 1, \frac{1}{2})$.

We conclude that, if $4m^2 \neq 1$, the space of positive energy solutions of (5.2) carries precisely the irreducible representation $D^S(m+1,\frac{1}{2})$. We have thus succeeded, in the case s = 0 and for $E_0 \neq 1$ and 2, in isolating each of the three irreducible representations of the induced representation

$$\mathrm{IND} \underset{sp(4,R)}{\overset{osp(4/1)}{\uparrow}} D(E_0,0) = D^S(E_0 - 1,0) \oplus D^S(E_0 - \tfrac{1}{2}, \tfrac{1}{2}) \oplus D^S(E_0,0) .$$

The three equations are, in the same order,

$$[\mathcal{D} - 2(E_0-1)]\, \Phi = 0 \quad ,$$

$$[\mathcal{M} + 4(E_0-1)(E_0-2)]\, \Phi = 0 \ ,$$

$$[\mathcal{D} + 2(E_0-2)]\, \Phi = 0 \quad .$$

The result is illustrated in Fig. 4.

The representation $D^S(m+1,\frac{1}{2})$ includes spins 0, $\frac{1}{2}$ and 1. The flat space analogue is the Wess-Zumino vector multiplet.[9,23] The spin-1 component is associated with the vector field (A_{ab}) or (A_α). Irreducibility implies that the equation for this vector field, contained in (5.2), must be precisely the spin-1 Fierz-Pauli equation in de Sitter space.[7] This has been meticulously double checked. The degree to which this equation is simplified and clarified by supersymmetry is really remarkable.

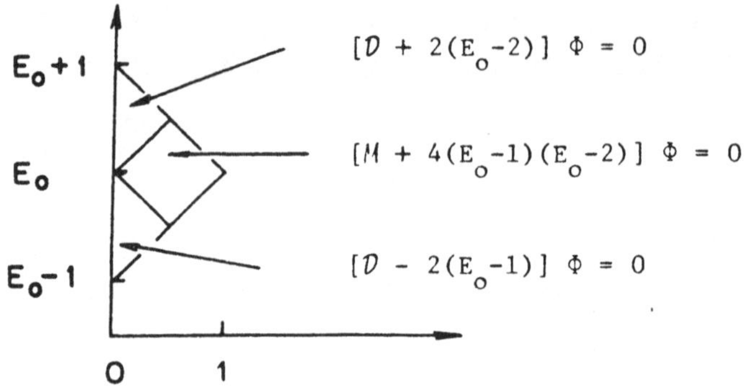

Fig. 4.

The wave equations that project out each of the three irreducible parts of the representation of osp(4/1) induced from the representation $D(E_0,0)$ of sp(4,R).

We next examine the exceptional cases $2m = \pm 1$. The wave equation (5.2) reduces to the "super de Sitter Maxwell equation":

$$\mathcal{M}\Phi = 0 \,, \tag{5.5}$$

and now we cannot use (5.1) to infer that $\mathcal{D}\Phi$ must vanish. Instead, we find a large subspace of solutions that we shall refer to as "gauge space" or as the space of gauge modes, namely

$$\text{Ker}\ \mathcal{M} \supset \text{Im}\ \mathcal{D} \equiv \text{gauge space} \,. \tag{5.6}$$

This subspace is invariant, but it is not invariantly complemented. However, any complement is contained in the generalized null space of $\mathcal{D}$, which is invariant. The identity (5.1) assures us that the minimal polynomial of $\mathcal{D}$ contains at most two factors $\mathcal{D}$. Therefore, any complement of the gauge space (5.6) is contained in the "mode space"

$$\text{Ker}\ \mathcal{D}^2 \equiv \text{mode space} \,. \tag{5.7}$$

This is the space that must be used for indefinite metric (Gupta-Bleuler) quantization. We shall now prove that, if a physical interpretation is possible at all, then the physical states of the theory are in the cohomology quotient:

$$\text{Ker}\ \mathcal{D}/[\text{Im}\ \mathcal{D} \cap \text{Ker}\ \mathcal{D}] = \text{physical space} \,. \tag{5.8}$$

The interpretation of (5.6) as an ignorable subspace implies that there exists an inner product $<\ ,\ >$ with respect to which the physical modes are orthogonal to the gauge modes:

$$<\Phi, \mathcal{D}\Psi> = 0\ , \quad \Phi \text{ physical}\ , \quad \Psi \text{ arbitrary}\ . \tag{5.9}$$

Invariance with respect to the superalgebra implies that $\mathcal{D}$ is formally selfadjoint and thus

$$<\mathcal{D}\Phi, \Psi> = 0\ . \tag{5.10}$$

The physical Hilbert space metric is based on < , >, after restriction and subsequent passage to the quotient by the radical. Hence (5.10) implies that $\mathcal{D}\Phi$ projects to zero in the physical Hilbert space. We can therefore impose the condition

$$\mathcal{D}\Phi = 0\ , \quad \text{(Lorentz condition)} \tag{5.11}$$

without losing any physical states.

This shows that, if a physical interpretation is possible, then the physical states are contained in (5.8). We now make a more detailed analysis to show that a physical interpretation is indeed possible, and that the space of physical states is precisely the cohomology quotient (5.8).

The Lorentz condition (5.11), together with the wave equation (5.5), tell us that

$$\mathcal{M}\Phi = -4(\square + 2)\ \Phi = 0$$

in the subspace of direct physical significance. As $\square + 2 = (E_0\text{-}1)(E_0 - 2)$ in $D(E_0,0)$, the inducing representation is D(2,0) or D(1,0) or a direct sum of these two. We know that the direct sum must be associated with two independent fields, and therefore deal with each irreducible inducing representation separately. A direct and elementary calculation shows that[17]

$$\mathrm{IND}\ \underset{sp(4,R)}{\overset{osp(4/1)}{\uparrow}}\ D(2,0) = D^S(1,0)$$

$$\oplus\ [D^S(2,0) \to D^S(\tfrac{3}{2},\tfrac{1}{2}) \to D^S(2,0)]\ . \quad (5.12)$$

The irreducible summand is associated with the eigenspace $\mathcal{D} = 2$, it is excluded by Eq. (5.7) and makes no contribution to the free quantum field operator. In the remaining, nondecomposable subspace the Lorentz condition (5.11) excludes the first $D^S(2,0)$, the quotient of "scalar" modes. This leaves the subspace Ker $\mathcal{D}$, in which the on-shell gauge modes, the second $D^S(2,0)$, form an invariant subspace and the quotient $D^S(\frac{3}{2},\frac{1}{2})$ is the unitary representation that is associated with the physical states.[18]

The other case,

$$\mathrm{IND}\ \underset{sp(4,R)}{\overset{osp(4/1)}{\uparrow}}\ D(1,0) = [D^S(\tfrac{1}{2},\tfrac{1}{2}) \to \{D^S(\tfrac{3}{2},\tfrac{1}{2}) \oplus \mathrm{Id}\} \to D^S(\tfrac{1}{2},\tfrac{1}{2})]$$

$$\oplus\ D^S(1,0)\ , \quad (5.13)$$

may be analyzed in parallel fashion. Note, however, the interesting appearance of the trivial representation Id in the physical subquotient.[17] In both cases it can easily be verified that the vector component of the superfield satisfies Maxwell's equation in de Sitter space. The invariant subspace represented by the last term within each nondecomposable triplet, $D^S(\frac{1}{2},\frac{1}{2})$ in (5.13) and $D^S(2,0)$ in (5.12), belongs to Im $\mathcal{D}$. The statement (5.8) has thus been fully justified.

The physical content of this de Sitter version of superelectrodynamics is revealed by [18]

$$D^S(\tfrac{3}{2},\tfrac{1}{2})\Big|_{sp(4,R)} = D(\tfrac{3}{2},\tfrac{1}{2}) \oplus D(2,1) ,$$

these being the massless representations with spins $\frac{1}{2}$ and 1.[7]

One can easily verify that

$$\mathcal{D}\, Q_\alpha\, \mathcal{D} = 0 .$$

It follows that

$$Q^\alpha\, \mathcal{D}\, Q_\alpha \propto \mathcal{M} .$$

The Lagrangian for Eq. (5.2) can therefore be written

$$\int d^4x\, d^4\theta\, (Q^\alpha Q^\beta\, \Phi)^\dagger\, (Q_\alpha Q_\beta \Phi) .$$

In this form the flat space limit can easily be compared with the conventional formulation.[20,22] A flat space discussion of the gauge subspace and the Lorentz condition may be found in Ref. 22.

6. The Simplest Superfield for N = 2 Supersymmetry

The operators L_{ab}, K_a, Q_a, Q_o given by Eqs. (1.2), (1.4), (3.5), and (3.4) satisfy the commutation relations of osp(4/2) and provide the simplest superfield realization. It is useful to understand how these formulas can be related to an induced representation.[13] Let $\mathfrak{g}$ denote the complex extension of osp(4/2). The even part $\mathfrak{g}_o$ of $\mathfrak{g} = \mathfrak{g}_o + \mathfrak{g}_1$ is sp(4,C) + C. The action of the even part on the odd part $\mathfrak{g}_1$ is the sum of two irreducible representations. The odd part thus splits

$$\mathfrak{g}_1 = \mathfrak{g}_1^{+} + \mathfrak{g}_1^{-} \quad ,$$
$$\mathfrak{g}_1^{\pm} = \text{Span}\,\{K_a \pm iQ_a\}\ . \tag{6.1}$$

The simplest superfield is obtained by inducing from the subalgebra

$$\mathfrak{g}_0 + \mathfrak{g}_1^{+} = b^{+}\ , \tag{6.2}$$

with $\mathfrak{g}_1^{+}$ acting trivially in the inducing representation. In this way one gets a superfield with only 4 Grassmann generators.

The simplest way to find the action of osp(4/2) on this simplest superfield is certainly not by explicitly evaluating the induced representation. A much simpler method was proposed long ago, in which osp(4/2) was interpreted as the superalgebra of conformal transformations of a Grassmann space with constant curvature.[21] An even simpler derivation may be found in the Appendix. When allowance is made for an arbitrary "conformal degree" one finds that

$$K_a = (1 - \theta^2/2i)\,\partial_a + i\theta^b\,L_{ab} - ic\theta_a\ , \tag{6.3}$$

$$Q_a = i(K_a - 2\partial_a) \quad , \tag{6.4}$$

$$Q_0 = 2(c - \theta\cdot\partial) \quad . \tag{6.5}$$

These expressions reduce to (1.4), (3.5) and (3.4) when c = 2. Unlike the situation for osp(4/1), different values of c give inequivalent representations of osp(4/2). See Appendix, Section 3.

The realization of b^+ from which the realization (6.3)-(6.5) is induced is

$$L_{ab} \to \kappa_{ab}\,, \quad Q_o \to 2c\,, \quad g_1^{\;+} \to 0\,. \tag{6.6}$$

In the induced representation L_{ab} is given by (1.2). Recall that we write L_{ab} instead of the more precise notation $\pi(L_{ab})$.

We shall now discuss the construction of invariant Lagrangians. The case N = 1 presented no difficulties, but for N = 2 the "central" charge Q_o gives rise to some complications. A kinetic action is an expression of the form

$$\int dx d\theta dx' d\theta' \; \Phi^{\dagger}(x,\theta)\; I(x,\theta;x',\theta')\; \Phi(x',\theta')\,. \tag{6.7}$$

Here $\Phi \to \Phi^{\dagger}$ denotes the involutive antiautomorphism of the Grassmann algebra that is generated by $i \to -i$ and $\theta_a \to \theta_a$. It is required that the action be invariant under osp(4/2), which means that the integral kernel must satisfy

$$\begin{aligned} &(K_a + K_a{}')^{\dagger}\; I(x,\theta;x',\theta') = 0\,, \quad a = 1,\ldots,4\,,\\ &(Q_\alpha - Q_\alpha{}')^{\dagger}\; I(x,\theta;x',\theta') = 0\,, \quad \alpha = 0,\ldots,4\,. \end{aligned} \tag{6.8}$$

The dependence on the parameter c can be factored out by setting

$$I = (1 + i\theta\cdot\theta' - \theta^2\,\theta'^2/4)^{4-c}\; J\,; \tag{6.9}$$

Eqs. (6.8) then reduce to the following conditions on J:

$$\begin{aligned} &(\theta\cdot\partial - \theta'\cdot\partial')\,J = 0\,, \quad (L_{ab} + L'_{ab})\,J = 0\,,\\ &\left[\partial_a + \frac{i}{2}\theta'^2\,\partial_a{}' + i\theta'^b\,L'_{ab}\right]\; J = 0 \end{aligned} \tag{6.10}$$

and the last equation with θ,θ' interchanged.

Invariant two-point functions are found in the same way. Invariance means that

$$<\Phi(x,\theta)\, \Phi^{\dagger}(x',\theta')> = \tilde{I}(x,\theta;x',\theta') \tag{6.11}$$

must satisfy the conditions (6.8), except that K_a and Q_α now appear instead of their formal adjoints. This simply means that the parameter c is replaced by 4-c. If we set

$$\tilde{I} = (1 + i\theta\cdot\theta' - \theta^2\, \theta'^2/4)^c\, \tilde{J}\,, \tag{6.12}$$

then the conditions on $\tilde{J}$ are precisely the same as the conditions on J; Eqs. (6.9).

Bilinear invariants of the form (6.7), and invariant two-point functions, are thus seen to exist. But (6.7) is not an acceptable Lagrangian in the usual sense unless J is the integral kernel of a differential operator. Now J is determined by its θ- and θ'-independent part J_o and the order of J as a differential operator is 4 + the order of J_o. Therefore, I is at best the integral kernel of a fourth order differential operator.

The realization of N = 2 supersymmetry in terms of unextended superfields has been discussed by Taylor.[3] The use of the covariant derivations to generate N = 2 extended Poincaré supersymmetry is the basis of Sokatchev's analysis.[24] The formulas (6.3)-(6.4) were first given in Ref. 21 (with c = 0).

7. Induction From An Irreducible Representation

Suppose now that the inducing representation is an irreducible, positive energy representation of b^+,

$$\pi_0 = D(E_0, s/z_0) , \tag{7.1}$$

where $Q_0 \to z_0$ and $g_1{}^+ \to 0$. Then, in the generic case, the induced representation is also irreducible:

$$\pi = \mathrm{IND} \underset{b+}{\overset{osp(4/2)}{\uparrow}} \pi_0 = D^s(E_0-1, s/z_0-1) \tag{7.2}$$

$$= D^s(E_0', s'/z_0') . \tag{7.3}$$

The representation (7.3) is unitarizable if[25]

$$E_0' - s' - 1 \geq |2z_0'| . \tag{7.4}$$

Below this limit it is unitarizable only if $s = 0$ and then only at the isolated points

$$E_0' = |2z_0'| \geq \tfrac{1}{2} , \tag{7.5}$$

$$E_0' = z_0' = 0 . \tag{7.6}$$

The induced representation (7.2) is in the continuous range of unitarizable representations if the parameters of π_0 satisfy

$$E_0 - s > 2z_0 \geq 2 . \tag{7.7}$$

We begin our discussion with this range.

In the case of strict inequality in (7.7) the induced representation (7.2) is irreducible. It follows that the only invariant wave operators are the fixed, θ-independent operators of the inducing representation. The second order Casimir operator, for instance, is trivial, as we now show. It is the operator

$$\mathscr{C}_2 = (1/4)(L_a{}^b L_b{}^a - i K^a K_a - i Q^a Q_a - 8Q_o{}^2) . \tag{7.8}$$

$$\text{In } D^s(E_o,s/z_o)\text{:}\ \ \mathscr{C}_2 = E_o(E_o-1) + s(s+1) - 2z_o{}^2 . \tag{7.9}$$

In the induced representation (7.2) the value is

$$E_o(E_o-3) + s(s+1) + 2 - 2(z_o-1)^2 ,$$

which is the same as the value of $\Box + 2 - 2(z_o-1)^2$, whence the operator identity

$$\mathscr{C}_2 = \Box + 2 - 2(z_o-1)^2 . \tag{7.10}$$

This identity may be used to derive (3.8) from (3.7). Note, however, that in Sections 2-5 we fixed $c = 2$ ($z_o = 1$). In the general case the correct formula is

$$\mathscr{D} = \frac{1}{2i} Q^\alpha Q_\alpha + (c-2)^2 .$$

We next investigate the limiting cases in which the induced representation is not irreducible. These are precisely the cases in which nontrivial, invariant wave equations exist. We shall see that the induced representation always remains indecomposable, even in these

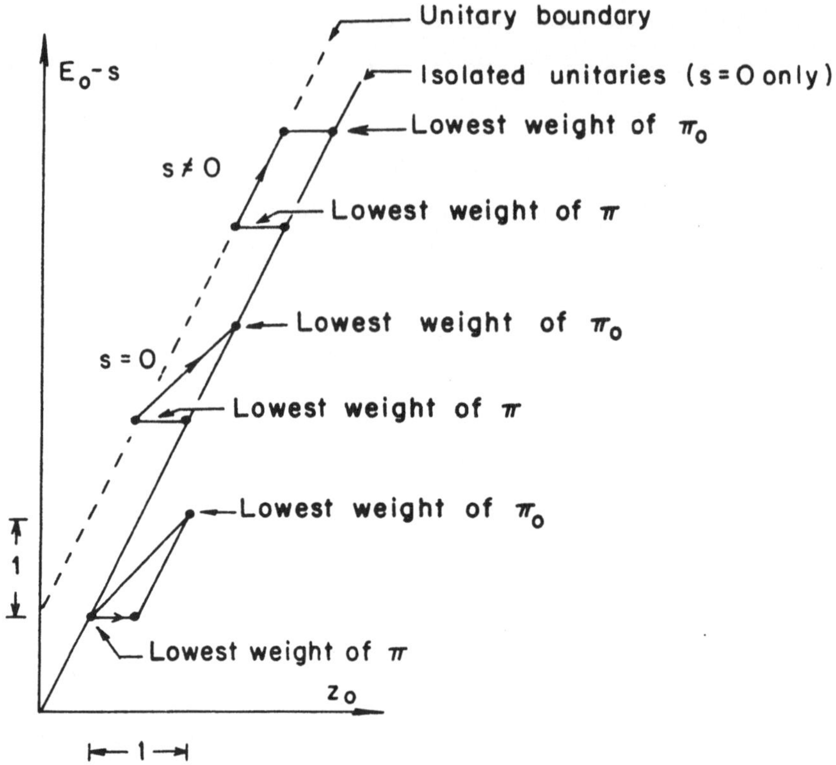

Fig. 5.

The Structure of representation π of osp(4/2) induced from the representation $\pi_0 = D(E_0, s/z_0)$ of b^+. Each dot represents an irreducible representation of osp(4/1).

special cases. The invariant wave equations characterize the irreducible, invariant subspaces, which allows us to find these equations without laborious calculations.

It will be convenient to limit ourselves at first to the case when only one type of reduction occurs. To be precise, we shall impose the restrictions that validate the generic formula (2.7) for the reduction of the induced representation on osp(4/1). The weights associated with the four components are

$$(E_o,s/z_o)$$

$$(E_o-\tfrac{1}{2},s-\tfrac{1}{2}/\,z_o-\tfrac{1}{2}) \quad (E_o-\tfrac{1}{2},s+\tfrac{1}{2}/\,z_o-\tfrac{1}{2}) \tag{7.11}$$

$$(E_o-1,s/z_o-1)$$

except that the one to the left is absent when $s = 0$. We are concerned with two special cases:

$$\text{(i)}\ \ E_o - s = 2z_o > 2\ , \tag{7.12}$$

$$\text{(ii)}\ \ E_o = 2z_o - 1 > 2\ ,\ \ s = 0\ . \tag{7.13}$$

In case (i) the induced representation lies at the end of the unitary range (7.4). In case (ii) it is at the isolated point (7.5). See Fig. 5.

The common feature that characterizes both of these cases is that two of the weights in (7.11) become Weyl equivalent. In case (i) we must treat the spinless case separately.

(ia) $E_o - s = 2z_o > 2$ and $s \neq 0$. The weights at the bottom and to the left in (7.11) are Weyl equivalent. [One verifies, in particular, that $\mathscr{C}_2$ takes the same value in these two representations.] The induced representation is now

$$D^s(2z_0+s-1, s/\ z_0-1) \to D^s(2z_0 + s - \tfrac{1}{2}, s - \tfrac{1}{2}/\ z_0 - \tfrac{1}{2}) . \tag{7.14}$$

(ib) $E_0 = 2z_0 > 2$ and $s = 0$. The weight to the left in (7.11) is absent and the weights at the top and at the bottom are Weyl equivalent, hence

$$\pi = D^s(2z_0-1, 0/\ z_0-1) \to D^s(2z_0, 0/\ z_0) . \tag{7.15}$$

(ii) $E_0 = 2z_0 - 1 > 2$ and $s = 0$. The weight at the bottom of (7.11) is Weyl equivalent to the one to the right, and

$$\pi = D^s(2z_0-2, 0/\ z_0-1) \to D^s(2z_0 - \tfrac{3}{2}, \tfrac{1}{2}/\ z_0 - \tfrac{1}{2}) . \tag{7.16}$$

From these results one easily extracts the following information about the reduction of irreducible representations of osp(4/2) on the subalgebra osp(4/1) $\oplus$ O(2). In the generic case there are 4 components, compare (7.11),

$$D^s(E_0,s/z_0)|osp(4/1) \oplus O(2)$$

$$= D^s(E_0+1,s)\cdot(z_0+1) \oplus D^s(E_0 + \tfrac{1}{2}, s - \tfrac{1}{2})\cdot(z_0 + \tfrac{1}{2})$$

$$\oplus D^s(E_0 + \tfrac{1}{2}, s + \tfrac{1}{2})\cdot(z_0 + \tfrac{1}{2}) \oplus D^s(E_0,s)\cdot(z_0) . \tag{7.17}$$

The second term is to be ignored when $s = 0$. If $E_0 = 2z_0 + 1 + s$ ($z_0 > 0$), then the first two terms drop out. If $s = 0$ and $E_0 = 2z_0 > 0$, then only the last term remains. In other words

$$D^S(2z_o+1+s,s/z_o)|osp(4/1) \oplus O(2)$$

$$= D^S(2z_o + \tfrac{3}{2} + s,\ s + \tfrac{1}{2})\cdot(z_o + \tfrac{1}{2}) \oplus D^S(2z_o + 1 + s,\ s)\cdot(z_o) \quad (7.18)$$

$$D^S(2z_o,0/z_o)|osp(4/1) \oplus 0(2) = D^S(2z_o,0)\cdot(z_o)\ . \quad (7.19)$$

In the special cases considered so far there is a relationship between E_o and z_o, but no upper bound on E_o. We can therefore perform a Wigner-Inonü contraction to massive representations of the N = 2 extended super Poincaré algebra. The result in each case is that the O(2) central charge is related to the mass. These special cases are thus directly related to the phenomenon of central charges in N = 2 extended super Poincaré field theories.

In contrast, the remaining special cases arise from special, low values of E_o. The contraction to the flat space limit gives either massless field theories or else vacuum representations with vanishing four momentum. We limit ourselves to s = 0 and the three special values of E_o in which the reduction formula (2.7) is replaced by (2.8), (2.9) or (2.10). These are limits of the cases (ib) and (ii) that were treated above, as E_o takes the values $\frac{3}{2}$, 2 and 1.

The complete list of these six cases follows.

(ib) $E_o = \frac{3}{2}$, $z_o = \frac{3}{4}$. Singleton.

$$\pi = D^S(\tfrac{3}{2},0/\tfrac{3}{4}) \to D^S(\tfrac{1}{2},0/-\tfrac{1}{4}) \to D^S(\tfrac{3}{4},0/\tfrac{3}{4})$$

(ib) $E_o = 2$, $z_o = 1$. Super QED.

$$\pi = D^S(2,0/-1) \to D^S(1,0/0) \to D^S(2,0/1)$$

(ib) $E_0 = 1,\ z_0 = \frac{1}{2}$.

$$\pi = D^S(0,0/-\tfrac{1}{2}) \to D^S(1,0/\tfrac{1}{2})$$

(ii) $E_0 = \frac{3}{2},\ z_0 = \frac{5}{4}$. Singleton.

$$D^S(1,\tfrac{1}{2}/\tfrac{3}{4}) \to D^S(\tfrac{1}{2},0/\tfrac{1}{4}) \to D^S(1,\tfrac{1}{2}/\tfrac{3}{4})$$

(ii) $E_0 = 2,\ z_0 = \frac{3}{2}$

$$\pi = D^S(1,0/\tfrac{1}{2}) \to D^S(\tfrac{3}{2},\tfrac{1}{2}/1)$$

(ii) $E_0 = 1,\ z_0 = 1$. Super QED.

$$\pi = D^S(\tfrac{1}{2},\tfrac{1}{2}/-\tfrac{1}{2}) \to [D^S(1,0/0) \oplus \mathrm{Id}] \to D^S(\tfrac{1}{2},\tfrac{1}{2}/\tfrac{1}{2})\ .$$

8. Wave Equations for N = 2

As pointed out above, no interesting wave equation exists in the generic case in which the induced representation is irreducible. We are here concerned with the special cases in which wave equations do exist.

We begin with the relatively simple situations that are related to the problem of central charges, Eqs. (7.15) and (7.16). In each of these cases the second member is realized in an invariant, irreducible subspace and it is enough to discover the operators that annihilate them. Now refer to Fig. 4. In (7.15) the invariant subspace carries just one irreducible representation of osp(4/1), it corresponds to the upper triangle in the figure, and the equation that defines it is

$$(\mathscr{D} + 4z_0 - 4)\,\Phi = 0\,. \tag{8.1}$$

This equation thus projects out the irreducible osp(4/2) representation $D^s(2z_0,0/z_0)$, and the restriction of this representation to osp(4/1) is the irreducible representation $D^s(2z_0,0)$ of that sub-superalgebra.

In the special cases of massless fields and singletons one can determine invariant equations for some or all of the invariant subspaces. However, such equations are Lorentz conditions rather than wave equations. Since the modules are indecomposable, the only true wave operators are the invariant operators of the inducing representation.

Eq. (8.1) is the only known instance of an osp(4/2) invariant wave equation for the case s = 0. We have verified that, if the wave equation $(\mathscr{D} - 2m)\Phi = 0$ is invariant, then necessarily $m = 2 - 2z_0$; see Appendix, Section 3.

The existence of an osp(4/2) invariant wave equation does not guarantee that one can find an invariant Lagrangian to go with Eq. (8.1). The reason for this is that the wave operator in (8.1) is not an osp(4/2) scalar, but a component of an invariant tensor. This situation is characteristic of very degenerate representations of Lie algebras as well as superalgebras. They are characterized, not by the values of the Casimir operators, but by much larger ideals in the enveloping algebras. Because of the very strong analogies that exist between conformal field theories and N = 2 field theories (the difficulties are quite the same in both cases), it is relevant to quote the large ideals associated with massless representations of the conformal group. They can be found in Ref. 26. In the case at hand one finds that (8.1) is a component of a tensorial equation of the form

$$(E_{\alpha\beta} - kN_{\alpha\beta})\,\Phi = 0 \,.$$

Here indices run over 6 values, with

$$E_{\alpha\beta} = N^{\gamma\delta}\, L_{\alpha\gamma} L_{\beta\delta} \pm (\alpha,\beta)$$

$$(L_{\alpha\beta}) = \begin{pmatrix} L_{ab}, & K_{aj} \\ K_{ib}, & K_{ij} \end{pmatrix}, \quad (N_{\alpha\beta}) = \begin{pmatrix} \eta_{ab} & 0 \\ 0 & \frac{1}{2i}\, g_{ij} \end{pmatrix},$$

and k is a constant. This leads to enormous complications, as can be seen from the construction of the relaxed hypermultiplet.[27]

9. The Spinor Superfield and de Sitter Chirality

The spinor superfield is a main ingredient of super Yang-Mills theories. It can be studied by two complementary methods, by induction or by tensor calculus. We begin with an incomplete account of the representation

$$\mathrm{IND}\ \underset{sp(4,R)}{\overset{osp(4/1)}{\uparrow}}\ D(E_0,\tfrac{1}{2}) \tag{9.1}$$

The reduction formula

$$\mathrm{IND} \uparrow D(E_0,\tfrac{1}{2}) = D^S(E_0-1,\tfrac{1}{2}) \oplus D^S(E_0-\tfrac{1}{2},0)$$

$$\oplus\, D^S(E_0-\tfrac{1}{2},1) \oplus D^S(E_0,\tfrac{1}{2}) \tag{9.2}$$

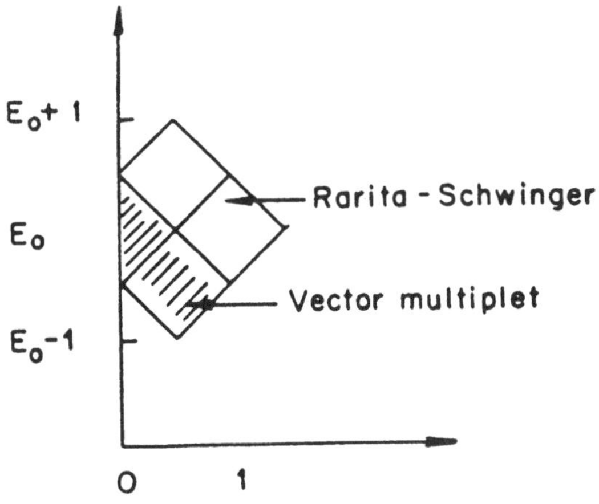

Fig. 6.

Content of spinor superfield. The representation induced from $D(E_0, \frac{1}{2})$ decomposes, for most values of E_0, in four components. The shaded area is the image of the gradient map. The spinor superfield contains more components, but these can be eliminated by transverse and divergence conditions and they do not appear in the induced representation.

is a special case of Eq. (2.7) and holds for all but a small set of special values of E_0. (We suppose that $E_0 > 0$.) The general case is illustrated in Fig. 6. The most interesting exceptions are $E_0 = \frac{5}{2}, \frac{3}{2}$ and $\frac{1}{2}$, for in these cases the physical representation $D^S(\frac{3}{2}, \frac{1}{2})$ of super QED appears. It is easy to see that

$$\mathrm{IND} \uparrow D(\tfrac{5}{2},\tfrac{1}{2}) = [D^S(2,0) \to D^S(\tfrac{3}{2},\tfrac{1}{2}) \to D^S(2,0)]$$

$$\oplus\ [D^S(\tfrac{5}{2},\tfrac{1}{2}) \to D^S(2,1) \to D^S(\tfrac{5}{2},\tfrac{1}{2})]\,. \tag{9.3}$$

The other special cases have not been analyzed.

Normally it is the highest spin that is physically relevant. A naive approach to gauge theory would suggest that the physically relevant part of (9.3) is $D^S(2,1)$, the spin content of which is given by

$$D^S(2,1)|sp(4,R) = D(2,1) \oplus D(\tfrac{5}{2},\tfrac{3}{2})\,. \tag{9.4}$$

But no local gauge theory with this spin content is known. In super Yang Mills theories it is the first part of (9.3) that is relevant, and this is the reason for the unconventional appearance of these theories.

The tensorial approach consists of multiplying the representation induced from $D(E_0,0)$ by a finite dimensional representation. The simplest example is the 5-spinor superfield

$$(A_\alpha) = (A_a, A_o)\,,$$

where a runs from 1 to 4. If (K_a) continues to denote the operators associated with the action of osp(4/1) on scalar superfields, then the operators (K_a') that define the action on spinor superfields is found by composition with the 5-dimensional representation. The prototype spinor superfield is the gradient

$$\Phi_a = Q_\alpha \Phi \, . \tag{9.5}$$

The commutation relations give

$$K_a{}' Q_b \Phi \equiv -Q_b K_a \Phi = K_a Q_b \Phi - \eta_{ab} Q_o \Phi \, ;$$

$$K_a{}' Q_o \Phi \equiv Q_o K_a \Phi = K_a Q_o \Phi + 2i \, Q_a \Phi \quad ,$$

and thus in general

$$K_a{}' A_b = K_a A_b - \eta_{ab} A_o \, ,$$

$$K_a{}' A_o = K_a A_o + 2i \, A_a \, . \tag{9.6}$$

Kinematical aspects of super QED can be described by a scalar superfield, as was seen in Section 5. Therefore, it can also be described by a spinor superfield of the type (9.5). But there are other ways. A spinor superfield that accounts for the states of free super QED is not necessarily the gradient of scalar superfield. This is important, for we would like to introduce interactions by means of the minimal substitution

$$Q_\alpha \rightarrow Q_\alpha - e\mathcal{A}_\alpha \tag{9.7}$$

in the matter Lagrangian. Since any gradient field can be removed by a gauge transformation on the matter fields, it is clear that super QED cannot be described by a gradient field within the context of minimal coupling.

An example of a spinor superfield that is not a gradient, but that is nevertheless capable of describing the states of noninteracting super QED, can be constructed as follows. We note that (y_{ab}) is an invariant sp(4,R) tensor. From it, we can construct an invariant osp(4/1) tensor $(Y_{\alpha\beta})$. This has been relegated to the Appendix, Section A4. We put

$$\mathcal{A}_\alpha = Y_\alpha{}^\beta \, Q_\beta \, \Phi \, . \tag{9.8}$$

This is not a gradient for all Φ. Instead we find that the field (9.8) is a gradient if and only if Φ belongs to Im $\mathfrak{D}$.

Recall that Im $\mathfrak{D}$ is the gauge subspace in the description of super QED by a scalar superfield. The formula (9.8) maps this subspace to what constitutes the gauge subspace in the description of super QED by a spinor superfield, and this spinorial gauge subspace is precisely the subspace of gradient superfields.

The Poincaré spinor superfield was first discussed by Adjei and Akyempong.[28] The gauge theory based on the flat space analog of (9.4) was proposed by Ogievetskii and Sokatchev.[29] The super QED particle content of the spinor superfield was discussed in detail by Salam and Strathdee in their 1978 review.[22] An approximate version of chirality on de Sitter space was discussed by Ivanov and Sorin.[12]

Acknowledgements

It is a pleasure to acknowledge helpful conversations with H. Araki, B. Binegar, R. Finkelstein, M. Flato and M. Villasante. I thank the National Science Foundation and the Japanese Ministry of Education for financial support.

APPENDICES

A1. Linear action for osp(2n/1).

The superalgebra osp(2n/N) is defined as follows. Let $W = W_o + W_1$ be a graded vector space, with even dimension N and odd dimension 2n. Let η be a non-degenerate antisymmetric form on W_1 and g a non-degenerate symmetric form on W_o. Let (ξ^a) a = 1,...,2n be coordinates for W_1 and (x^j) j = 1,...,N, coordinates for W_o and define

$$\xi_b = \xi^a \eta_{ab} , \quad x_j = x^i g_{ij} ,$$

$$\partial_a = \partial/\partial\theta^a , \quad \partial_j = \partial/\partial x^j .$$

Now osp(2n/N) is the superalgebra of infinitesimal endomorphisms of W that preserve the form $\eta \oplus g/2i$. It is convenient to promote W_1 to a Grassmann algebra,

$$\xi^a \xi^b = - \xi^b \xi^a , \quad \partial_a \xi^b = \delta_a^{\ b} - \xi^b \partial_a ;$$

then osp(2n/N) is the superalgebra of first order differentiable operators that annihilate the polynomial

$$\xi^a \xi_a + x^j x_j/2i \,. \tag{A.1}$$

The even part consists of infinitesimal orthogonal transformations of W_o and symplectic transformations of W_1. The operators

$$K_{aj} = x_j \partial_a - 2i\xi_a \partial_j \,, \quad a = 1,\ldots,2n \;;\; j = 1,\ldots,N \,, \tag{A.2}$$

form a basis for the odd part.

In the case N = 1 we set $g_{11} = 1$, $x^1 = x_1 = x$, and restrict ourselves to the "hyperboloid"

$$\xi^a \xi_a + x^2/2i = 1 \,. \tag{A.3}$$

We introduce homogeneous coordinates

$$\lambda^a = \xi^a/x \,,$$

in terms of which

$$K_{a1} = \partial_a + 2i\lambda_a \lambda\cdot\partial \,, \quad \partial_a = \partial/\partial\lambda^a \,.$$

To obtain the operators K_a used in the text we finally introduce θ^a by

$$\lambda^a = \theta^a/(1 + \theta^2/2i) \,,$$

to obtain

$$K_{a1} = (1 + \theta^2/2i)\,\partial_a + i\theta_a \theta\cdot\partial \,, \quad \partial_a = \partial/\partial\theta^a \,. \tag{A.4}$$

This agrees with Eq. (1.3) in the case when $\kappa_{ab} = 0$; that is, when the action on the space time manifold is ignored. [Concerning the term $-2i\theta_a$ in Eq. (1.4), see below.]

For the extension of this result to the general case covered by Eqs. (1.2) and (1.4) we have two methods. The original idea was to look at M as a fiber over the Grassmann algebra. The coordinates (y^α) of the de Sitter hyperboloid are related to the antisymmetric tensor with components

$$y_{ab} = y^\alpha (\gamma_\alpha)_{ab}$$

(as in Section 4). This is provisionally interpreted as a tensor field over the Grassmann algebra, and the operators K_a act accordingly, as Lie derivatives. The coefficients of K_a are, by (A.4),

$$K_a^{\ b} = (1 + \theta^2/2i)\, \delta_a^{\ b} + i\, \theta_a \theta^b \,.$$

The variation of y_{bc} induced by K_a is

$$\delta y_{bc} = (\partial_b K_a^{\ d})\, y_{dc} - (b,c) + 2i\theta_a y_{bc}$$

$$= i\theta^d\, (\eta_{bd} y_{ac} + \eta_{ba} y_{dc} - b,c) \,. \qquad \text{(A.5)}$$

Of course, this is not really a tangent vector to de Sitter space, but an action on fields on that space, valued in the Grassmann algebra, is nevertheless defined:

$$K_a = K_{a1} - \tfrac{1}{2}\, \delta y_{bc} \partial^{bc} = K_{a1} + i\, \theta^b\, \kappa_{ab} \,.$$

This agrees with Eq. (1.3).

We can add a term $-ic\theta_a$, with constant c, without spoiling the commutation relations. Representations that differ only in the choice of c are equivalent. The intertwining operator is constructed in Section 3 of this Appendix.

This calculation justifies the formulas of Section 1 in the case when the operators κ_{ab} are the vector fields associated with the action of sp(4,R) on de Sitter space. The method can evidently be generalized, but the deepest way to understand Eqs. (1.2) and (1.3) is to derive them as the action of osp(2n/1) in an induced reprsentation.[13] Nevertheless, it must be pointed out that the evaluation of the formulas of the induced action is rather complicated for large superalgebras. For osp(2n/1) it has been done both ways, and the calculation presented here is decidedly the simpler.

A2. Linear action for osp(2n/2).

Let us return to Eqs. (A.1) and (A.2). In the case N = 2 it is convienent to use circular (complex) coordinates for 2-space, in which

$$g_{11} = g_{22} = 0 , \quad g_{12} = g_{21} = 1 .$$

The projective cone

$$2i\xi^a \xi_a + x^j x_j = 0 , \quad (\xi^a, x^j) \simeq (\lambda\xi^a, \lambda x^j) \tag{A.6}$$

can then be identified with the Grassmann algebra, just as Dirac's projective 6-cone is identified with compactified Minkowski space. Projective coordinates (θ^a) a = 1,...,2n are defined by

$$\xi^a = x^1 \theta^a ,$$

x^2 is eliminated by means of Eq. (A.6),

$$x^2 = - ix^1 \theta^a \theta_a .$$

Restriction to homogeneous functions,

$$(\xi^a \partial_a + x^j \partial_j) \psi(\xi,x) = c\psi(\xi,x) , \tag{A.7}$$

permits a factor $(x^1)^c$ to be factored out, leaving in its memory a multiplier.

The action of (A.2) then takes the form

$$K_{a1} = -i\theta^2 \partial_a + 2i\theta_a(\theta\cdot\partial - c) ,$$

$$K_{a2} = \partial_a = \partial/\partial\theta^a \quad . \tag{A.8}$$

The generalization to include the action on de Sitter space proceeds exactly as in the case N = 1. The result is

$$K_{a1} = i\theta^2 \partial_a + 2i\theta^b L_{ab} - 2ic\theta_a ,$$

while K_{a2} is unaffected. The operators (6.3), (6.4) are the following linear combinations

$$K_a = \tfrac{1}{2} K_{a1} + K_{a2} \quad ,$$

$$Q_a = i(\tfrac{1}{2} K_{a1} - K_{a2}) \quad . \tag{A.9}$$

A3. Intertwining operators.

The formula

$$(1 - \theta^2/2i)^{1-n}\,\partial_a(1 - \theta^2/2i)^n = (1 - \theta^2/2i)\,\partial_a + in\theta_a$$

shows us how to intertwine two realizations of osp(4/1) that differ only by the choice of the parameter c in Eq. (6.3). If

$$K_a^{(c)} = (1 - \theta^2/2i)\,\partial_a + i\theta^b L_{ab} - ic\theta_a\,,$$

then

$$(1 - \theta^2/2i)^{-n} K_a^{(c)}(1 - \theta^2/2i)^n = K_a^{(c-n)}\,.$$

Taking $n = c-2$ we transform (6.3) to the form (1.4) that was used in Sections 1-5. We need to know the expression $Q_o^{(2)}$ for Q_o in the same realization:

$$Q_o^{(2)} = -2(1 - \theta^2/2i)^{2-c}\,(\theta\cdot\partial - c)\,(1 - \theta^2/2i)^{c-2}$$

$$= -2[\theta\cdot\partial - c + i(c-2)\,\theta^2\,(1 + \theta^2/2i)] \qquad \text{(A.10)}$$

We shall use this formula to determine the values of m for which the wave equation (4.2) is osp(4/2) invariant. We use the notation of Eq. (3.3):

$$\Phi = \phi + \theta\cdot\psi + \ldots \qquad , \qquad \text{(A.11)}$$

$$-\tfrac{1}{2}\,Q_o^{(2)}\Phi = \phi' + \theta\cdot\psi' + \ldots \quad . \qquad \text{(A.12)}$$

Suppose that $(\mathcal{D} - 2m)\, \Phi = 0$. When $s = 0$, then this is the same as imposing Eqs. (4.3)-(4.5) on the coefficients $\phi, \psi, \ldots$. According to (A.10), the coefficients of the field (A.12) are

$$\phi' = -c\phi \ , \quad \psi' = (1-c)\psi \ , \quad A' = (2-c)(A-\phi) \ ,$$

$$\chi' = (3-c)\chi + (c-2)\psi \ , \quad F' = (4-c)F + (2-c)(A-\phi/2) \ .$$

Invariance of the wave equation under osp(4/2) is the requirement that, whenever (4.3)-(4.5) hold, then the coefficients ϕ', ψ', ... satisfy the same equations. One easily verifies that this happens if and only if $m = 2 - c$. In the flat space limit this means that the central charge is equal to the mass.

A4. Invariant fields.

It was seen, in Sections A.1 and A.2, that "θ-space" can usefully be thought of as a conic in a superspace of higher dimension. Here we use the term superspace in the sense of the graded vector space of the defining representation of osp(2n/N). This space is given a natural structure of superalgebra, and the "conics" are ideals in this superalgebra.

It is fruitful to push the formal analogy with real conics a step further. Consider the imbedding of de Sitter space in R^5. It gives rise to an invariant, R^5-valued function on de Sitter space, namely the imbedding function $x \to y$. Hence x is a point in de Sitter space and $y = (y^\alpha)$ $\alpha = 0,1,2,3,5$ is the corresponding point in R^5. This function is invariant if R^5 is considered as an so(3,2) module. If ℓ_{ab} are the matrices of the natural action of so(3,2) in R^5, and κ_{ab} are the vector fields of the action of so(3,2) in de Sitter space, then

$$\kappa_{ab} y^{\alpha}(x) - (\ell_{ab})_{\beta}{}^{\alpha}\, y^{\beta}(x) = 0\,.$$

Hence y is an invariant vector field, or more precisely an invariant vector valued function, to be compared with the Kroenecker symbol.

To define an analogous invariant vector field in 5-dimensional superspace, we first nail down a standard form of the natural action of osp(4/1) in a 4+1-dimensional graded vector space. The prototype will be obtained from Eq. (3.6):

$$[K_a, Q_b] = \eta_{ab} Q_0 \quad ,$$

$$[K_a, Q_0] = -2iQ_a \quad .$$

Hence

$$(Q^{\alpha}) = (Q^a, Q^0)$$

will be a standard multiplet. Another standard multiplet is given by (ξ^a, x), in the notation of Section A1. In terms of θ^a, this multiplet is

$$(\Theta^{\alpha}) = (\Theta^a, \Theta^0) \qquad ,$$

$$\Theta^a = \theta^a / f_- = \eta^{ab}\, \Theta_b \,,$$

$$\Theta^0 = f_+ / f_- = \Theta_0 \qquad ,$$

$$f_{\pm} = 1 \pm \theta^2 / 2i \qquad .$$

This is our invariant, vector valued function. We have

$$K_a \Theta_b = \eta_{ab} \Theta_o \,, \quad K_a \Theta_o = -2i\Theta_a \,.$$

Some simple formulas follow:

$$\Theta^\alpha \Theta_\alpha = \Theta^a \Theta_a + \frac{1}{2i} \Theta^o \Theta_o = \frac{1}{2i} \,,$$

$$\Theta^\alpha Q_\alpha = \Theta^a Q_a + \frac{1}{2i} \Theta^o Q_o = -2i \,,$$

$$Q^\alpha Q_\alpha = Q^a Q_a + \frac{1}{2i} Q^o Q_o = 2i\mathcal{D} \,.$$

A transverse gradient is defined by

$$\tilde{Q}_\alpha \Phi = (Q_\alpha - 4\Theta_\alpha)\, \Phi \,,$$

$$\Theta^\alpha \tilde{Q}_\alpha = 0 \qquad ;$$

it is useful in the formulation of chirality.

Recall that $y = (y_{ab})$ can be interpreted as a tensor field on θ-space. It is the projection on the supercone of an invariant bi-5-spinor field $Y = (Y_{\alpha\beta})$. With the help of (A.5) we find that the components are

$$Y_{ab} = y_{ab} - i(\theta_a \theta^d \, y_{db} - a,b) \,,$$

$$Y_{ao} = -2i \, f_- \, \theta^d \, y_{da} \qquad ,$$

$$Y_{oo} = 4\theta^c \, \theta^d \, y_{cd} \qquad .$$

This tensor is transverse, and traceless:

$$\Theta^{\alpha} Y_{\alpha\beta} = 0 , \quad Y^{\alpha}{}_{\alpha} = 0 .$$

The square is proportional to the transverse projection operator,

$$Y_{\alpha}{}^{\beta} Y_{\beta}{}^{\gamma} = -y^2 (\delta_{\alpha}{}^{\gamma} - 2i\Theta_{\alpha}\Theta^{\gamma}) ,$$

where

$$(\delta_{\alpha}{}^{\gamma}) = \begin{pmatrix} \delta_a{}^c & 0 \\ 0 & 2i \end{pmatrix} .$$

The de Sitter curvature constant

$$y^2 = -\frac{1}{4} y^{ab} y_{ab} = 1/\rho$$

has been normalized to unity. In this section we have used Eq. (6.3) with c = 2.

References.

1. P. Breitenlohner and D. Z. Freedman, Ann. Phys. 144, 249 (1982).
2. S. J. Avis, C. J. Isham, and D. Storey, Phys. Rev. D18, 3565 (1978).
3. J. G. Taylor, Nucl. Phys. B169, 484 (1980).
4. S. Okubo, Phys. Rev. D29, 269 (1984) and D29, 1865(E).
5. D. Z. Freedman, Phys. Rev. D15, 1173 (1977); A. Das and D. Z. Freed Nucl. Phys. B120, 221 (1977); P. K. Townsend, Phys. Rev. D10, 280 (1977); S. W. McDowell and F. Mansouri, Phys. Rev. Let. 38. 739 (1977); P. K. Townsend and P. van Nieuwenhuizen, Phys. Lett. 67B, 439 (1977); S. Deser and B. Zumino, Phys. Rev. Lett. 38, 1433 (197 A. H. Chamseddine, Nucl. Phys. B131, 494 (1977); E. Cremmer and B. Julia, Phys. Lett. 80B, 48 (1978) and Nucl. Phys. 159B, 141

(1979); M. J. Duff and C. N. Pope, in "Supergravity '82" S. Ferrara, J. G. Taylor, and P. van Nieuwenhuizen, eds.), World Scientific; F. Englert, Phys. Lett. 119B, 339 (1980).

6. M. Flato and C. Fronsdal, Lett. Math. Phys. 2, 421 (1978); Phys. Lett. 97B, 236 (1980) and J. Math. Phys. 22, 1100 (1981).
7. C. Fronsdal, Phys. Rev. D12, 3819 (1975).
8. B. Binegar, C. Fronsdal, and W. Heidenreich, Ann. Phys. 149, 254 (1983).
9. Yu. A. Gol'fand and E. P. Likhtman, JETP Lett. 13, 323 (1971).
10. A. Salam and J. Strathdee, Nucl. Phys. 80B, 499 (1974).
11. P. H. Dondi and M. Sohnius, Nucl. Phys. B81, 317 (1974); B. L. Aneva, S. G. Mikhov and D. Ts. Stoyanov, Theor. Math. Phys. 27, 502 (1976) and 31, 177 (1977).
12. B. W. Keck, J. Phys. A8, 1819 (1975); B. Zumino, Nucl. Phys. B127, 189 (1977); E. A. Ivanov and A. S. Sorin, J. Phys. A13, 1159 (1980).
13. C. Fronsdal and T. Hirai, "Unitary Representations of Supergroups," UCLA preprint, April 1985, preceding paper.
14. E. Angelopoulos, M. Flato, C. Fronsdal, and D. Sternheimer, Phys. Rev. D23, 1278 (1981).
15. W. Heidenreich, Phys. Lett. 110B, 461 (1982).
16. H. Araki, Commun. Math. Phys. 97, 149 (1985).
17. M. Flato and C. Fronsdal, "Spontaneously Generated Field Theories, Zero-Center Modules, Colored Singletons and the Virtues of N = 6 Supergravity," UCLA preprint TEP/20, December 1984, following paper.
18. P. Breitenlohner and D. Z. Freedman, Ann. Phys. 144, 249 (1982).
19. L. O'Raifeartaigh, Nucl. Phys. B89, 418 (1975).
20. S. Ferrara, J. Wess and B. Zumino, Phys. Lett. B51, 239 (1974); A. Salam and J. Strathdee, Phys. Lett. B51, 353 (1974).
21. C. Fronsdal, Lett. Math. Phys. 1, 165 (1976).
22. A. Salam and J. Strathdee, Fortschr. d. Physik 26, 57 (1978).

23. J. Wess and B. Zumino, Nucl. Phys. B70, 39 (1974).

24. E. Sokatchev, Nucl. Phys. B99, 96 (1975).

25. D. Z. Freedman and H. Nicolai, Nucl. Phys. B237, 342 (1984); L. Castell, T. Künemund and W. Heidenreich, Starnberg Preprint; H. Nicolai, CERN preprint TH. 3882, 1984.

26. B. Binegar, C. Fronsdal, and W. Heidenreich, J. Math. Phys. 24, 2828 (1983).

27. P. S. Howe, K. S. Stelle, and P. K. Townsend, Nucl. Phys. B124, 519 (1983).

28. S. A. Adjei and D. A. Akyempong, Nuovo Cim. 26A, 84 (1975); S. J. Gates, Jr., Phys. Rev. D16, 1727 (1977); W. Siegel, Phys. Lett. 85B, 333 (1979).

29. V. I. Ogievetskii and E. Sokatchev, JETP Lett. 23, 58 (1976); S. J. Gates, Jr. and R. Grimm, preprint 1984.

SPONTANEOUSLY GENERATED FIELD THEORIES, ZERO-CENTER MODULES, COLORED SINGLETONS AND THE VIRTUES OF N = 6 SUPERGRAVITY

by

M. Flato and C. Fronsdal

ABSTRACT. Attention is called to an interesting property of the space of one-particle states in some especially important massless field theories: the appearance of a one-particle ghost with zero energy. It appears in conformal as well as de Sitter electrodynamics, in the physical sectors of the field mode representations of the respective symmetry groups. It appears again in super de Sitter electrodynamics based on the superalgebra osp(4/1) and in super conformal electrodynamics based on su(2,2/1). We next construct two families of extended super QED, incorporating this property, based on osp(4/N) and on su(2,2/N). There is precisely one osp(4/N) theory and one su(2,2/N) theory of this type for each value of N. The osp(4/6) theory is the same as N = 6 extended supergravity, this is the only one among this family of osp(4/N) theories in which the highest spin is 2. All the one particle states are massless, and in the osp(4/N) theories they can be interpreted as states of two colored singletons. We also critically examine the concept of the Witten index in flat space as well as in de Sitter supersymmetric field theories.

C. Fronsdal (ed.), Essays on Supersymmetry, 123–162.

0. Introduction

This paper attempts to develop the idea that the most significant field theories are those that are created spontaneously. We shall describe an interesting property that characterizes electromagnetism, try to relate the existence of this phenomenon to "spontaneous creation," and determine a select family of field theories with the same property. These include conformal and supersymmetric generalizations of electrodynamics, and N = 6 extended supergravity. (Incidentally, we describe supersymmetric field theories by techniques that seem to be simpler than those found in the literature.) Proposed applications include super electroweak theories that are significantly different from usual Yang-Mills theories (some but not all of the vector fields are in the adjoint representation of the internal group) and unifications of super electroweak with supergravity. Some of these theories seem especially well suited for incorporation of the Higgs-Kibble mechanism, but their most important characteristic is the natural inclusion of a degenerate vacuum. Preon schemes based on colored singletons also integrate naturally with this approach.

The concept of spontaneous creation of field theories is one that we advance cautiously and without any pretense of precision. It is based on the following more or less connected reflections. (i) A spin lattice exhibits unexpected degrees of freedom, distinct from those of the individual spin excitations. Phonons and solitons appear "spontaneously" because of the existence of configurations with arbitrarily small excitation energy. (ii) Like most students of electromagnetism at one time or another, we have toyed with the idea that the electromagnetic field might be replaced by a nonlocal expression involving the current. The analogy with spin waves is striking. The free Dirac electron field theory is invariant under constant phase transformations. Almost constant phase transformations may be compared with the gentle deformations of the spin lattice ground state that explain the phonons, and related to the existence of low energy photons. (iii) Induced gravity[1] also embodies the idea of field modes created spontaneously, and some preon models (for massless gauge fields)[2] have much the same concept incorporated into them. Note, however, that ordinary gravity and simple supergravity do not fit into our scheme (though N = 6 extended supergravity does fit); this shows that it is more selective than, though compatible with, the basic idea underlying induced gravity.

We shall now draw attention to an interesting property of electromagnetism that seems especially relevant for understanding what it is that characterizes spontaneously created field modes. This property also relates strongly to associated ideas about spontaneous symmetry breaking. It is not expected that spontaneously created field modes will have mass. Created from nothing, they should have some of the characteristics of the vacuum. Some quantum numbers should be zero--but which ones? What is it that distinguishes quantum numbers such as mass, which photons have not, from energy and momentum and spin, which they do have? The answer must be related to the circumstance that we are here concerned with phenomena engendered by symmetry. Most relevant for QED is perhaps its highest symmetry--Poincaré symmetry or even conformal invariance. Now we observe that free photons are described by representations of the Poincaré group in which all the Casimir operators vanish, as they do for the vacuum (the trivial representation). So we arrive at the first, and only a very preliminary, formulation of the main premise of this paper; namely that the central role of electrodynamics has something to do with the fact that the photon representations of the Poincaré group have the same values of the Casimir operators as the trivial representation. A much more precise characterization of electrodynamics (and its generalizations) will come below.

The values of the Casimir operators are enough to characterize the representations of compact groups, but far from sufficient to fix representations of the Poincaré group. All massless representations, of whatever helicity, have the same values of both Casimir operators. Yet we do not believe that spinless particles, say, belong to the select family of particles that are spontaneously generated. What, then, is unique to the photon representation? This is a question that we cannot answer in this narrow context, but a very satisfactory reply can be given after either extending the Poincaré group to the conformal group, or else replacing it by the 3+2 de Sitter group. By so doing we shall discover the degenerate vacuum advertised above, and the property that we believe characterizes spontaneously created field theories. Additional impetus for admitting a (small but non-zero) constant curvature comes from the very efficient infrared regularization that accompanies it, and the clarification that this brings to discussions of symmetry breaking. This point will be discussed in some detail near the end of this introduction, where the problem posed by the Poincaré group will also be taken up.

The irreducible, positive energy representations of the 3+2 de Sitter group are fully characterized by the lowest value E_o of the energy and the spin s,[3] and they are thus denoted without ambiguity as $D(E_o,s)$.[5] The massless representations[4] are D(s+1,s) and the Dirac singletons (the constituents or "square roots" of the massless particles[6,7]) are D(1/2,0) and D(1,1/2). Among the massless representations there is just one, D(2,1), that has both Casimir operators equal to zero. The irreducible, positive energy representations of the conformal group are fully characterized by the minimal weight E_o,j_1,j_2, where E_o is the smallest value of the conformal energy and j_1,j_2 are the so(4) indices of the ground states.[8] The massless representations D(s+1,s,0) and D(s+1,0,s) have the unique property of remaining irreducible when restricted to the Poincaré or de Sitter subgroups.[9] Among them only D(2,1,0) and D(2,0,1) have the property that all the Casimir operators vanish. We shall refer to these exceptional representations, D(2,1) of so(3,2) and D(2,1,0) and D(2,0,1) of the conformal group, as "zero-center modules," since they are characterized by the fact that the centers of the enveloping algebras are zero. These are precisely the representations associated with the physical states of de Sitter[4] and conformal[10] QED, except that an additional vacuum mode also appears. In fact, the significance of zero-center is precisely the fact that such a zero energy mode occurs. It is this vacuum mode that we feel is associated with spontaneous creation of field theories, and with spontaneous symmetry breaking as well. We must now explain how the vacuum mode intrudes into the space of physical states.

In gauge theories the space of field modes is larger than the space of physical states. Thus, in Poincaré QED, the space of field modes includes scalar, transverse and longitudinal (gauge) photons. All these modes contribute to a non-decomposable Poincaré module[11] that we represent as (see also Fig. 1).

$$D(0,0) \rightarrow [D(0,1) \oplus D(0,-1)] \rightarrow D(0,0) \,.$$

The invariant subspace of gauge modes appears on the right, it carries the irreducible representation with zero mass and zero helicity. The arrow[10] (distinct from the sign used for direct sums) is meant to indicate that this invariant subspace is not invariantly complemented. The physical helicities ±1 are in the middle; with the gauge modes they make up the submodule that satisfies the Lorentz condition. As indicated, this subspace also is not invariantly complemented. As one knows, the

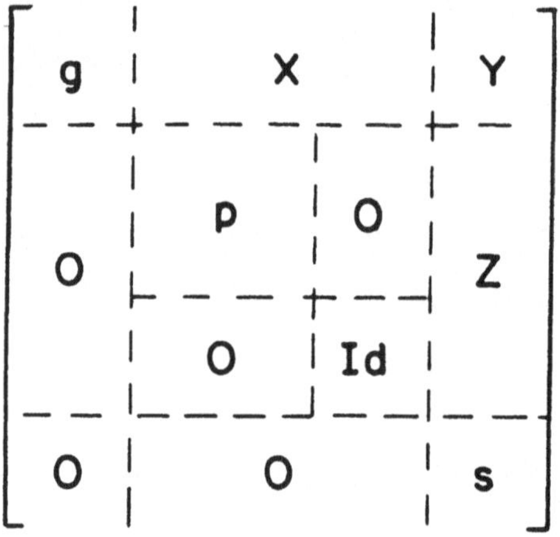

Fig. 1.

General structure of Gupta-Bleuler nondecomposable representation. The physical representation p is unitary; the gauge and scalar representations g and s must be conjugate to each other in the sense of Ref. 12 but need not be unitary. The operator Id is the identity operator, it can appear only in the semi-simple case, and only in zero-center modules. In the case of conformal gravity a nonunitary ghost appears in the physical sector, that must be eliminated by a constraint.

Lorentz condition is incompatible with the free field commutation relations; this is simply because the scalar modes are needed as momenta canonically conjugate to the gauge modes. The entire non-decomposable module is a Gupta-Bleuler triplet--it has the crucial property of supporting a non-degenerate symplectic structure.[12]

In de Sitter electrodynamics the situation is similar except for a small but, we now believe, essential difference. The total field module is the direct sum of two parts,[13]

$$D(3,0) \rightarrow D(2,1) \rightarrow D(3,0)$$

and

$$D(1,1) \rightarrow \begin{bmatrix} D(2,1) & \\ & \oplus\ \mathrm{Id} \end{bmatrix} \rightarrow D(1,1)\ .$$

The physical representation D(2,1) of the 3+2 de Sitter group contracts, in the limit of zero curvature, to either of the two representations D(0,±1) of the Poincaré group.[14] Both triplets are essential for a smooth limit,[13] but that is not the point on which we want to elaborate here. Instead, we wish to emphasize the unexpected intrusion of the trivial representation Id = D(0,0), the vacuum mode. In conformal QED (whether Minkowski or de Sitter does not matter) one finds the following field model Gupta-Bleuler triplet:[10]

$$D(1,1/2,1/2) \rightarrow \begin{bmatrix} D(2,1,0) \oplus D(2,0,1) \\ \oplus\ \mathrm{Id} \end{bmatrix} \rightarrow D(1,1/2,1/2)\ .$$

Once again the vacuum mode shows up. This phenomenon is repeated in super de Sitter QED and in super conformal QED, where the trivial representations of the respective superalgebras appear.

The significance of the defining property of irreducible, zero-center modules, the vanishing of all the Casimir operators, is now evident. The irreducible subquotients of any non-decomposable representation, of any Lie algebra or superalgebra, must have the same infinitesimal character.[15] Thus, if a nondecomposable representation contains the trivial representation, then all the other irreducible subquotients must be zero-center modules. Any representation that contains a finite number of irreducible subquotients, all of them zero-center, will be called a zero-center representation. In such representations all the Casimir operators are nilpotent. Zero-center modules are of great interest in mathematics,[16] and they are the point of departure of the Zuckerman translation functor.[17] This is a prescription that can be used to "translate" the electromagnetic field mode representations to corresponding gravitational field mode representations. It has been applied with good effect by Binegar[18] and it appears to have validity for superalgebras as well. We shall not use the translation functor in this paper, but we call attention to the very distinguished role that is played by electrodynamics and by zero-center modules in general.

As we mentioned, super electrodynamics also contains vacuum modes. Both super de Sitter QED based on osp(4/1) and super conformal QED based on su(2,2/1) are formulated in terms of zero-center representations of the respective superalgebras, as is shown in detail in Sections 3 and 5. In Sections 4 and 6 we turn this observation into a constructive principle, obtaining two families of extended QED's based on osp(4/N) and su(2,2/N). For $n \leq 4$ the highest spin is 1 and these theories are well known super-Yang-Mills theories. For $N = 5$ the highest spin is 3/2; this theory is of a new type and it has not yet been constructed explicitly. For $N = 6$ the highest spin is 2 and the osp(4/6) theory is $N = 6$ extended supergravity. Of all the extended de Sitter supergravities, only this one associates a zero-center representation with the free quantum field. We propose that it may be the one of greatest physical significance, since it is the one that seems most likely to be created spontaneously, and the only one that contains the interesting vacuum mode with its hint of spontaneous symmetry breaking. This theory has a photon in the 1-dimensional representation of so(6), and an additional set of spin 1 particles in the adjoint representation. The spin 0 particles are also in the adjoint representation and seem destined to play the role of Higgsons in the massification of the spin 1 particles, leaving only the photon massless.

For applications to super electroweak theory the su(2,2/N) family seems very promising. In this paper we limit ourselves to establishing the existence of this family of theories, in Section 6. It needs to be stressed that the (super) conformal theories described here are unitary; unlike some of those found in the literature.[19] See in this connection Refs. 18 and 20.

Once again, we call attention to the Dirac singletons,[6,7] and to their generalizations, the "colored singleton" representations of osp(4/N).[21] In this paper they are used only to construct the massless representations (by reducing the direct product of two singleton representations), but their importance as physical constituents of massless particles[6,7] far transcends their mathematical utility. A very simple and beautiful construction of the osp(4/N) singletons is given in Section 4.

Next, a comment on the formulation of electrodynamics in dimensions other than 4. Conventional three-dimensional electrodynamics is infrared nonrenormalizable. Perhaps a more reasonable version can be found by applying the ideas of this paper in a three-dimensional space time. Work along these lines is in progress. We believe that electrodynamics, in any dimension, should have 1/r static potentials, and we expect that this is what we shall find in a theory based on a zero-center field representation.

The fact that the idea of zero-center modules has been formulated more precisely, and hence more profitably, in de Sitter space than in Minkowski space should not be taken as the final word on the subject. It is true that there is no unitary, irreducible representation of the Poincaré group that can be extended by the trivial representation,[22] and that for this reason the interesting vacuum mode cannot appear in the non-decomposable Poincaré field module. But this just means that the question of what is the pertinent analogue of zero-center modules of semi-simple groups remains open. Since the problem is closely related to infrared regularization, one should perhaps investigate the Fock representation instead of the one-particle representation. Another possibility, also suggested by quantum field theory, is to look for extensions of massless representations by an infinite sum of identity representations. In the meantime we may work in de Sitter space (with arbitrarily small curvature) as long as it is convenient, and pass to the flat space limit if and when this becomes preferable. The conformal group furnishes another alternative. One can also argue that the vacuum mode is a strong indication of the relevance of the semi-simple groups for the description of massless fields.

We return to the discussion of infrared regularization in de Sitter space. The low energy spectra of massless field theories are illustrated in Fig. 2. In Fig. 2a we show the continuous spectrum of field theories in flat space, and in Fig. 2b we see the effect of a finite volume cut-off. It should be noted that a finite volume only guarantees discretization of the momentum, not necessarily of the energy, and that this may invalidate[23] an attempt[24] to count the degeneracy of zero-energy modes. In de Sitter space, the effect of finite curvature is seen in Fig. 2c; here the energy spectrum is always discrete. [It is possible to

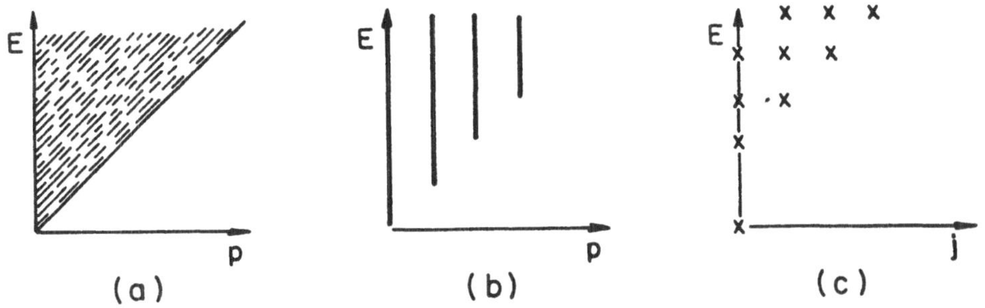

Fig. 2. Low energy spectrum

(a) Continuum in flat space theories.

(b) Discretization of the momentum spectrum introduced in flat space by finite volume cut-off.

(c) Discrete spectrum of the energy characteristic of massless particles of field theories in de Sitter space.

produce a partly continuous spectrum by taking a direct integral of irreducible representations with $E_o > 1$, but there will always be a gap between 0 and 1/2. Note that the energy is measured in units of the square root of the curvature constant.]

In a flat space supersymmetric field theory the number of fermion states is equal to the number of boson states, for every value of the four-momentum, except possibly at vanishing four-momentum. The difference $N_f - N_b$ between the numbers of fermion and boson vacuum states is the Witten index[24] of the theory. As far as we know, an analogous result has not been proved for super de Sitter theories. A precise rule, found to hold in numerous cases but not proved in general, is as follows. First define the "degree" of any positive energy, irreducible representation $D(E_o,s)$ of so(3,2). The number of states with energy $E = E_o + n$, n a positive integer, grows as $kn^2/2$. The coefficient k is an integer and will be referred to as the degree of the representation. Here is the value of k in all important cases:

Massive case, $E_o > s + 1\ ,\ \ k = 2s + 1\ ,$

Massless case, $E_o = s + 1 > 1\ ,\ \ k = 2\ ,$

$D(1,0)$ and $D(2,0)\ ,\ \ k = 1\ ,$

$D(\tfrac{1}{2},\tfrac{1}{2})\ ,\ \ k = 2\ ;\quad D(1,1)\ ,\ \ k = 1\ .$

Finite dimensional representations and singletons have degree zero. What we have found is that in positive energy representations of osp(4/N) the sum of the degrees of the integral spin representations of SO(3,2) is always equal to the sum of the degrees of the half-integral spin representations. In a superconformal theory involving only massless representations we get the same result with each massless, irreducible representation having degree 1, which is consistent with the usual flat space result for massless particles. We now elaborate on this point.

In the limit of flat space, a slight refinement of this result tells us that the number of boson states at momentum p is the same as the number of fermion states at the same momentum, as long as $p \neq 0$. No information is obtained about states at or near zero energy, however, because in de Sitter supersymmetry the low lying boson and fermion states are not matched. This may be surprising, since the introduction of a small, constant curvature may be regarded as an infrared regularization device similar to a volume cut-off, and Witten's analysis[24] would at first sight seem to apply. The explanation lies in the fact that the anticommutation relations of the supersymmetry algebra are affected by the curvature. Indeed, some further reflection on this problem soon makes one realize that the details of the volume cut-off (the shape of the walls, for example) materially affect the low lying spectrum. Making the box very large simply shifts the lack of control to a region close to zero energy. Thus we conclude that there is no way that one can learn anything about the number of states at or near zero energy by means of a volume cut-off. This agrees with other results[23] that also indicate that the Witten index is highly regularization-dependent.

Finally, let us explain our attitude towards the vacuum mode that characterizes zero-center field theories. Actually, this mode is a ghost in conformal QED, and perhaps always. Therefore, let us first take a defensive attitude. The appearance of a finite representation in the physical sector is a very common phenomenon; it is only in the case of zero-center modules that it is one-dimensional and unitary. Therefore, zero-center field theories are better than other field theories. We know that the vacuum mode seems to be harmless in conformal QED. It may be that the finite modes can always be decoupled; nevertheless, zero-center modules seem to have the best chance to make it. However, our attitude to the vacuum mode is quite different from this, and not at all defensive. We believe on the contrary that the zero energy ghost may play an important role as a natural regulator of the infrared problem, and that the degeneracy of the vacuum is relevant to spontaneous symmetry breaking at the tree level. The fact that this ghost appears in de Sitter field theories, without the direct intervention of conformal invariance, is one more reason for persisting in our attitude towards the cosmological constant: physics has much to gain, and nothing to lose by admitting that it can never be shown experimentally to be exactly zero.

1. De Sitter Electrodynamics

De Sitter space is a 4-dimensional pseudo-Riemannian space; the signature of the metric $(g_{\mu\nu})$ is +--- and the curvature tensor has the form

$$R_{\mu\nu\lambda\rho} = \rho(g_{\mu\lambda}g_{\nu\rho} - g_{\mu\rho}g_{\nu\lambda}), \quad \rho = \text{const.} > 0 .$$

Locally, de Sitter space can be imbedded isometrically in 3+2 dimensional pseudo-Euclidean space and identified with the hyperboloid

$$u^2 \equiv u_0^2 - u_1^2 - u_2^2 - u_3^2 + u_5^2 = 1/\rho . \tag{1.1}$$

Details of this mapping have been published many times.

Maxwell's equations, in any pseudo-Riemannian space, in the absence of sources, are

$$\partial_\mu(-g)^{1/2} g^{\mu\nu} g^{\rho\lambda} (\partial_\nu A_\lambda - \partial_\lambda A_\nu) = 0 . \tag{1.2}$$

We focus on the case when $(g_{\mu\nu})$ is the de Sitter metric; the indices μ,ν,λ take the values 0,...,3. The imbedding into R^5 leads to the equivalent equation[25,13]

$$\text{tr.pr.}[(\Box+2)\mathcal{A}_\alpha + (1/\rho)\, \partial_\alpha \partial\cdot\mathcal{A}] = 0 , \tag{1.3}$$

where $u\cdot\mathcal{A}(u) = 0$, indices α,β,γ take the values 0,...,3,5,

$$\Box = -u^2\partial^2 + u\cdot\partial(u\cdot\partial + 3) ,$$

and "tr.pr." means transverse projection relative to u.

The action of SO(3,2) on the vector potential is

$$(T_\Lambda\mathcal{A})_\alpha (u) = \Lambda_\alpha{}^\beta\, \mathcal{A}_\beta(\Lambda^{-1}u) , \quad \Lambda \in SO(3,2) . \tag{1.4}$$

The infinitesimal action is given by operators $L_{\alpha\beta} = -L_{\beta\alpha}$ defined by

$$(L_{\alpha\beta}\mathcal{A})_\gamma (u) = u_\alpha \partial_\beta \mathcal{A}_\gamma(u) + \delta_{\alpha\gamma}\mathcal{A}_\beta(u) - (\alpha,\beta) . \tag{1.5}$$

The modes of the free quantum field are the Fourier components of the homogeneous propagator (two-point function), and depend on the choice of gauge. The gauge is fixed by changing the coefficient of the second term in (1.2). The modified form of (1.3) is

$$\text{tr.pr.}[(\Box+2)\mathcal{A}_\alpha + (c/\rho)\, \partial_\alpha \partial\cdot\mathcal{A}] = 0 .$$

The condition $u\cdot\mathcal{A} = 0$ is retained. Other forms of gauge fixing, that relax this restriction,[26] offer some advantages, but they will not be reviewed here.

The two-point function is taken to be SO(3,2) invariant, and its Fourier components form an so(3,2) module for a non-decomposable, positive energy (extremal weight, K-finite) representation. It turns out that there are two possibilities. When $c = 2/3$ they are[13]

$$D(3,0) \to D(2,1) \to D(3,0) \quad . \tag{1.6}$$

$$D(1,1) \to [D(2,1) \oplus \mathrm{Id}] \to D(1,1) \; . \tag{1.7}$$

Both are relevant; electrodynamics in de Sitter space can be built up on either (1.6) or (1.7), but both are needed to secure a smooth flat space limit.[13] When $c \neq 2/3$ (and $c \neq 1$) the Gupta-Bleuler triplets are[27]

$$\begin{array}{cccc} D(3,0) \to & D(2,1) & \searrow & \\ & & & D(3,0) \\ & D(3,0) & \nearrow & \end{array}$$

and

$$\begin{array}{ccccc} & \nearrow & D(2,1) & \searrow & \\ D(1,1) & & & & D(1,1) \\ & \searrow & \mathrm{Id} & \nearrow & \\ & \nearrow & & & \\ D(1,1) & & & & \end{array}$$

The notation was explained in the introduction. All the extensions are non-trivial, and the inclusion of the trivial representation $\mathrm{Id} = D(0,0)$ shows that all the representations are zero-center modules. Note that the triplets associated with different values of the gauge fixing parameter c are inequivalent.

2. Conformal Electrodynamics

The conformal extension of electrodynamics is based on the existence of unitary, irreducible representations of SO(4,2) that remain irreducible when restricted to either of the subgroups associated with ordinary electrodynamics: Poincaré and de Sitter. They are the positive energy, zero-center modules D(2,1,0) and D(2,0,1).[9] The restrictions to the Poincaré group are D(0,1) and D(0,-1), respectively, while

$$D(2,1,0)|so(3,2) = D(2,1) = D(2,0,1)|so(3,2) . \tag{2.1}$$

The positive energy, K-finite, irreducible, zero-center representations form one Weyl equivalence class,[10] and include D(4,0,0), $D(3,\frac{1}{2},\frac{1}{2})$, $D(1,\frac{1}{2},\frac{1}{2})$ and Id = D(0,0,0), besides the physical representations D(2,1,0) and D(2,0,1).

The simplest formulation of conformal electrodynamics makes use of Dirac's realization[28] of compactified Minkowski space as the projective 6-cone

$$y^2 \equiv \delta^{\alpha\beta}\, y_\alpha y_\beta \equiv y_0^{\,2} - y_1^{\,2} - y_2^{\,2} - y_3^{\,2} - y_4^{\,2} + y_5^{\,2} = 0 .$$

$$\lambda y_a \simeq y_a , \quad \lambda \neq 0 .$$

Indices, α,β,γ in this section take the values 0,...,5. The free field equations, with the simplest choice of gauge fixing, are

$$\delta^{\alpha\beta}\, \partial_\alpha \partial_\beta a_\gamma(y) = 0 .$$

The action of SO(4,2) on the 6-cone and on the 6-potential is similar to the action of SO(3,2) given by Eqs. (1.4) and (1.5). The solutions of the free field equations include a space of positive energy modes that carry the following K-finite, non-decomposable representation of so(4,2):[10]

$$D(1,\tfrac{1}{2},\tfrac{1}{2}) \to [D(2,1,0) \oplus D(2,0,1) \oplus \mathrm{Id}] \to D(1,\tfrac{1}{2},\tfrac{1}{2}) \,. \qquad (2.2)$$

Other choices of gauge fixing lead to inequivalent modules.

The restriction of (2.2) to so(3,2) is the direct sum of (1.6), (1.7) and two copies each of D(1,0) and D(2,0).[13] The latter appear in the gauge and scalar sectors only,

$$D(1,\tfrac{1}{2},\tfrac{1}{2})|\mathrm{so}(3,2) \;=\; D(1,0) \oplus D(2,0) \oplus D(1,1) \,.$$

3. De Sitter Super Electrodynamics

The analogue, in 3+2 de Sitter space, of the super Poincaré algebra of Wess and Zumino,[29] is the orthosymplectic superalgebra osp(4/1).[30] The even part of osp(4/1) is the real symplectic algebra in 4 dimensions, isomorphic to so(3,2). To be definite, let η be a non-degenerate symplectic form on R^4, and identify the even part of osp(4/1) with sp(η).

Zero-center osp(4/1) modules can be constructed in terms of scalar superfields. A scalar superfield is a polynomial $\Phi(\theta)$ in 4 anticommuting quantities (θ^a) a = 1,...,4, with coefficients that are fields on de Sitter space. More precisely, the θ's are the generators of a Grassmann algebra (to which we adjoin a unit), and Φ is a field on de Sitter space valued in this algebra.[31] An action on Φ of the even subalgebra sp(η) is defined by the operators

$$L_{ab} = i\kappa_{ab} + is_{ab}\,, \quad a,b = 1,\ldots,4\,, \tag{3.1}$$

where

$$s_{ab} = -\,\theta_a\partial_b - \theta_b\partial_a\,,$$
$$\theta_b = \theta^a\,\eta_{ab}\,, \quad \partial_a = \partial/\partial\theta^a\,, \tag{3.2}$$

and (κ_{ab}) are the vector fields associated with the action of sp(η) on de Sitter space. If (γ_α) $\alpha = 0,1,2,3,5$ is a set of real, 4-by-4 matrices satisfying

$$\gamma_\alpha\gamma_\beta + \gamma_\beta\gamma_\alpha = -2\delta_{\alpha\beta}\,,$$

then in the notation of Section 1,

$$\kappa_{ab} = -(\gamma{\cdot}u\ \gamma{\cdot}\partial + u{\cdot}\partial)_a{}^c\,\eta_{cb}\,. \tag{3.3}$$

The operators (3.1), together with the odd operators

$$K_a = \theta^b\,L_{ab} + (1 - \theta^2/2i)\,\partial_a - 2i\theta_a$$
$$= i\theta^b\,\kappa_{ab} + (1 + \theta^2/2i)\,\partial_a + i\theta_a(\theta{\cdot}\partial - 2)\,, \tag{3.4}$$

define a realization of osp(4/1).[31] We have written θ^2 for $\theta^a\,\theta_a$.

To understand this construction it should be noticed that it is a special case of an induced representation.[30,32,33] Indeed, let D be any representation of sp(η) by operators κ_{ab} acting in some space V_{in}, and let Φ be a superfield with coefficients in V_{in}. [The Grassmann algebra S_1 generated by the θ's, with unit added, can be identified with the symmetric algebra of the odd part of osp(4/1), and then Φ belongs to $S_1 \otimes V_{in}$.] The realization given by (3.1) and (3.4) is simply the representation of osp(4/1) induced by the representation D of sp(η).[33]

Now let each component of Φ carry an irreducible representation D of sp(η), the action being given by the operators κ_{ab}. Then the action (3.1), (3.4) of osp(4/1) on Φ defines an induced representation D_s of the superalgebra. One sure way to generate zero-center modules is to choose D so that the lowest weight of D_s is zero. Since sp(η) is isomorphic to so(3,2), the lowest weight modules $D(E_o,s)$ are defined as in Section 1. The lowest weight of the representation induced by $D(E_o,s)$ is (E_o-1,s), so our first choice for D is the massless, spinless representation D(1,0). The other massless, spinless representation D(2,0) also leads to zero-center modules. In either case each component of Φ satisfies the wave equation for massless, spinless fields on de Sitter space, namely

$$(\Box + 2)\Phi = 0 \,, \tag{3.5}$$

$$\Box = \frac{1}{4}\kappa_a{}^b \kappa_b{}^a \,. \tag{3.6}$$

The operator $\Box$ is the same here as in Section 1.

It is an elementary exercise to evaluate the induced representations. The result is

$$\mathrm{IND}\overset{osp(4/1)}{\underset{so(3,2)}{\uparrow}} D(1,0) = D_s(1,0) \tag{3.7}$$

$$\oplus \{D_s(\tfrac{1}{2},\tfrac{1}{2}) \rightarrow [D_s(\tfrac{3}{2},\tfrac{1}{2}) \oplus \mathrm{Id}] \rightarrow D_s(\tfrac{1}{2},\tfrac{1}{2})\}$$

$$\mathrm{IND}\underset{so(3,2)}{\overset{osp(4/1)}{\uparrow}} D(2,0) = D_s(1,0) \tag{3.8}$$

$$\oplus \{D_s(2,0) \to D_s(\tfrac{3}{2},\tfrac{1}{2}) \to D_s(2,0)\} .$$

Here $D_s(E_0,s)$ is an irreducible representation of osp(4/1) with lowest energy E_0 and spin s. The reductions on the even part are

$$D_s(\tfrac{3}{2},\tfrac{1}{2})|so(3,2) = D(\tfrac{3}{2},\tfrac{1}{2}) \oplus D(2,1)$$

$$D_s(\tfrac{1}{2},\tfrac{1}{2})|so(3,2) = D(\tfrac{1}{2},\tfrac{1}{2}) \oplus D(1,0) \oplus D(1,1)$$

$$D_s(1,0)|so(3,2) = D(1,0) \oplus D(\tfrac{3}{2},\tfrac{1}{2}) \oplus D(2,0)$$

$$D_s(2,0)|so(3,2) = D(2,0) \oplus D(\tfrac{5}{2},\tfrac{1}{2}) \oplus D(3,0) . \tag{3.9}$$

The first of these osp(4/1) representations contains the massless representations of so(3,2) with spins $\frac{1}{2}$ and 1, first identified in Ref. 4. The third occurs in matter coupled supergravity;[34] this is the only one that is not a zero-center module of osp(4/1).

We have thus discovered two zero-center, super Gupta-Bleuler triplets, with the same physical modes but with different gauge and scalar modes. Both are realized on a scalar superfield, with each component satisfying the scalar wave equation (3.5). Both contain $D_s(1,0)$ as an extra direct summand that can be eliminated by restriction to the generalized null-space of the second order Casimir operator

$$\mathscr{C}_1 = \frac{1}{4} [\eta^{ac}\, \eta^{bd}\, L_{ab} L_{cd} + i\, \eta^{ab}\, K_a K_b] . \tag{3.10}$$

On $D_s(E_0,s)$ the value is

$$\mathscr{C}_1(E_o,s) = E_o(E_o - 2) + s(s + 1) .$$

On irreducible zero-center modules $\mathscr{C}_1$ vanishes, and the Gupta-Bleuler triplets are characterized by

$$\mathscr{C}_1{}^2\Phi = 0 .$$

The Lorentz condition, satisfied by the physical states and the gauge modes, is

$$\mathscr{C}_1\Phi = 0 ,$$

and the space of gauge modes is the subspace Im $\mathscr{C}_1$. The space of physical states is thus Ker $\mathscr{C}_1$/Im $\mathscr{C}_1$.

We do not give the details of the calculations that lead to (3.7) and (3.8), or to similar results given below. However, for those who are interested in computational methods we make an exception in Section 6. See also Ref. 10 where very similar calculations were given in complete detail.

4. Extended de Sitter Super Electrodynamics

We next consider the zero-center modules of the superalgebras osp(4/N). The even part of osp(4/N) is so(3,2) ⊗ so(N). Lowest weight, irreducible, K-finite osp(4/N) modules are labelled by the lowest weights (E_o,s/w), where w is a highest weight of so(N).[35] We shall give an explicit construction of zero-center Gupta-Bleuler triplets for the case N = 2, and return to the general case below.

It is easy to construct the Gupta-Bleuler triplets for N = 2. Consider again the superfield Φ of Section 3, with the wave equation (3.5). The superfield becomes an osp(4/2) module if we define[31] the action of the additional generators as

$$Q_a = iK_a - 2i\partial_a\,, \quad \text{odd generators, } a = 1,\ldots,4\,, \tag{4.1}$$

$$Z = \tfrac{1}{2}\,(\theta\cdot\partial - 2)\,, \quad \text{so(2) generator}\,. \tag{4.2}$$

This set of operators is the analogue, in de Sitter space, of the spinorial covariant derivative introduced by Ferrara, Wess and Zumino.[36] The second order Casimir operator

$$\mathcal{C}_2 = \frac{1}{4}\,[\eta^{ac}\,\eta^{bd}\,L_{ab}L_{cd} + i\eta^{ab}\,K_aK_b + i\eta^{ab}\,Q_aQ_b - 8Z^2]\,,$$

applied to Φ gives identically

$$\mathcal{C}_2\Phi = (\Box + 2)\Phi\,.$$

In contrast with the situation with osp(4/1), the space of solutions of the wave equation is here an ordinary null-space of $\mathcal{C}_2$.

A straightforward calculation gives

$$\mathrm{IND}\underset{so(3,2)\,\otimes\,so(2)}{\overset{osp(4/2)}{\uparrow}}[D(1,0)\otimes Id]$$

$$= D_s(\tfrac{1}{2},\tfrac{1}{2}/1) \to [D_s(1,0/0) \oplus Id] \to D_s(\tfrac{1}{2},\tfrac{1}{2}/-1) \tag{4.3}$$

$$\mathrm{IND}_{\,so(3,2)\,\otimes\,so(2)}^{\;osp(4/2)}\ [D(2,0)\otimes \mathrm{Id}]$$

$$= D_s(2,0/2) \to D_s(1,0/0) \to D_s(2,0/-2)\ . \tag{4.4}$$

The scalar and gauge modules remain irreducible when restricted to osp(4/1), while the osp(4/2) module $D_s(1,0/0)$ reduces to the direct sum of two osp(4/1) modules, $D_s(1,0)$ and $D_s(\frac{3}{2},\frac{1}{2})$. The reduction on so(3,2) can be inferred from (3.9).

In the general case ($N \geq 2$) we begin by identifying some of the zero-center modules of osp(4/N). The second order Casimir operator, on the irreducible representation with lowest weight $(E_o,s/w)$, takes the value

$$\mathscr{C}_N(E_o,s/w) = E_o(E_o + N - 3) + s(s + 1) - \tfrac{1}{2}\, c(w)\ . \tag{4.5}$$

Here c(w) is the value of the second order Casimir operator of so(N) in the representation with highest weight w. The above value of $\mathscr{C}_N$ vanishes in the following cases (among others)

(i) $E_o = s = w = 0$

(ii) $E_o = s = \frac{1}{2}\ ,\ \ w = \overline{\omega}_1 = (1,0,0,...) \equiv \mathrm{vec}\ ,$

(iii) $E_o = 1\ ,\ \ s = 0\ ,\ \ w = \overline{\omega}_2 = (1,1,0,...) \equiv \mathrm{adj}\ .$

Here $\{\overline{\omega}_i\}$ i = 1,...[N/2] are the fundamental weights of SO(N).[37] The three SO(N) representations listed are the trivial representation, the N-dimensional defining (vector) representation and the adjoint representation. The list is not exhaustive, of course. By studying the super Verma module generated by the trivial K-type we discover the existence of non-decomposable representations

$$\mathrm{Id} \to D_S(\tfrac{1}{2},\tfrac{1}{2}/\mathrm{vec}) \quad \text{and} \quad D_S(\tfrac{1}{2},\tfrac{1}{2}/\mathrm{vec}) \to D_S(1,0)/\mathrm{adj}) \ ,$$

in which both extensions are non-trivial. This shows that these representations have zero center. On this basis we conjecture the existence of a Gupta-Bleuler triplet

$$D_S(\tfrac{1}{2},\tfrac{1}{2}/\mathrm{vec}) \to [D_S(1,0/\mathrm{adj}) \otimes \mathrm{Id}] \to D_S(\tfrac{1}{2},\tfrac{1}{2}/\mathrm{vec}) \ . \qquad (4.6)$$

This generalizes the triplet (4.3) that we constructed for the case N = 2. The other triplet (4.4) is also expected to have a generalization to osp(4/N).

Though (4.6) has not yet been concretely realized, we can easily deduce the most important properties of the physical component $D_S(1,0/\mathrm{adj})$. We begin with the observation that

$$D_S(1,0/\mathrm{adj}) \subset D_S(\tfrac{1}{2},0/\mathrm{spin}) \otimes D_S(\tfrac{1}{2},0/\mathrm{spin}) \ . \qquad (4.7)$$

Here $D_S(\tfrac{1}{2},0/\mathrm{spin})$ is the "colored singleton" (colored Dirac supermultiplet) representation, a very special and highly singular (in fact, the most singular) representation of osp(4/N). We need to describe this interesting representation before returning to discuss the inclusion (4.7).

The oscillator representation[38] of sp(4,R) is the direct sum

$$D(\tfrac{1}{2},0) \oplus D(1,\tfrac{1}{2}) = \mathrm{Rac} \oplus \mathrm{Di} \ .$$

The two components are the most singular representations of sp(4,R), characterized by the highest possible degeneracy (lowest Gel'fand-Kirillov dimension). In these representations the energy and the angular momentum are linked by

$$E = j + \tfrac{1}{2} . \tag{4.8}$$

The oscillator representation has a natural extension to osp(4/1), the Dirac supermultiplet $D_s(\frac{1}{2},0)$,[39] with

$$D_s(\tfrac{1}{2},0)|sp(4,R) = \mathrm{Rac} \oplus \mathrm{Di} . \tag{4.9}$$

Let (K_a, L_{ab}) be a basis for this representation of osp(4/1), satisfying[39]

$$K_a K_b = L_{ab} - \frac{i}{2}\eta_{ab} , \quad a,b = 1,\ldots,4 . \tag{4.10}$$

This includes the anticommutation relations that belong to the structure of osp(4/1), as well as the commutation relations that characterize the oscillator representation.

The internal group so(N) also has a very exceptional representation, the spin representation with highest weight $\overline{\omega}_{[N/2]} = (\frac{1}{2},\frac{1}{2},\ldots,\frac{1}{2})$. It has a natural extension to the Clifford algebra of order N, with

$$\gamma^j\gamma^k = \delta^{jk} + 2iM^{jk} , \quad j,k = 1,\ldots,N , \tag{4.11}$$

where $(M^{kj} = -M^{jk})$ is a basis for the spin representation of so(N). It is reducible if N is even.

Now let K_a, γ^i be the operators of the oscillator representation of osp(4/1), respectively the Clifford algebra (4.11), in spaces U and V, and let $K_a{}^i$ be the operators in $U \otimes V$ given by

$$K_a{}^i = K_a \otimes \gamma^i . \tag{4.12}$$

Then, if we identify the $K_a{}^i$ with the odd generators of osp(4/N), and $L_{ab} \otimes \mathrm{Id}$ and $\mathrm{Id} \otimes M^{ij}$ with the even generators L_{ab} and M^{ij}, we obtain an explicit realization of osp(4/N) belonging to the equivalence class $D_s(\frac{1}{2},0/\mathrm{spin})$. This construction gives us the remarkable reduction formula

$$D_s(\tfrac{1}{2},0/\mathrm{spin})|sp(4,R) \otimes so(N) = [D(\tfrac{1}{2},0) \oplus D(1,\tfrac{1}{2})] \otimes \mathrm{spin} . \tag{4.13}$$

Now we can return to (4.7).

Since the adjoint representation of so(N) is contained in spin $\otimes$ spin, it is evident that the physical representation $D_s(1,0/\mathrm{adj})$ is contained in the invariant subspace of the tensor product (4.7) that is generated from the states of lowest energy. Since the reduction of the direct product of two Dirac supermultiplets, as so(3,2) modules, namely[6]

$$[D(\tfrac{1}{2},0) \oplus D(1,\tfrac{1}{2})] \otimes [D(\tfrac{1}{2},0) \oplus D(1,\tfrac{1}{2})]$$

$$= D(1,0) \oplus D(2,0) \oplus 2 \sum_{s=1/2}^{\infty} D(s+1,s) ,$$

contains nothing but massless representations of so(3,2), it follows that the same is true of $D_s(1,0/\mathrm{adj})$; that is, all the physical states of the zero-center Gupta-Bleuler triplet are massless.

This approach to the study of the massless representations of osp(4/N) is due to Castell, Heidenreich and Künemund,[21] who also established the unitary limits. The construction of the singleton representation given above is a special application of the methods of Günaydin et al.[40] The representations of osp(4/3) were studied by Freedman and Nicolai,[35] and the general case of osp(4/N) by Nicolai.[41] Colored singletons were investigated by Nicolai and Sezgin, and by Castell, Heidenreich and Künemund.[21] The physical role of colored singletons as constituents of massless quarks and of gluons was first suggested in Ref. 7.

Now it is not difficult to find the reduction of the physical zero-center module $D_s(1,0/\text{adj})$ on the even subalgebra. We summarize the physical content of each of the 7 zero-center theories that have spins not exceeding 2:

spin:	0	$\frac{1}{2}$	1	$\frac{3}{2}$	2	
N =0			1			QED
1		1	1			SQED
2	1	1+1	1			SQED
3	3	1+3	1			SY-M
4	3	4	1			SY-M
5	10	5+10	1+5	1		SR-S
6	15	6+10+10	1+15	6	1	SG

The last is N = 6 supergravity, and this is the only de Sitter supergravity with a zero-center representation. Each theory contains a singlet spin-1 (photon). In the case of N = 6 the other spin-1 particles (vector mesons) are in the adjoint representation, accompanied by scalars (Higgsons). Note that the number of spin-0 particles was defined as the multiplicity of $D(1,0) \oplus D(2,0)$.

We want to emphasize what we feel is an excellent justification for going as far as to osp(4/6) in the choice of symmetry algebra. The restriction of the zero-center osp(4/2) module (4.3) to osp(4/1) is the reducible representation (3.7). Only the second, indecomposable part is a zero-center osp(4/1) module. The irreducible summand $D_s(1,0)$ is not zero-center. This is a simple example of a general phenomenon that should be noted. Enlargement of the symmetry algebra may lead to the inclusion, in a zero-center module, of a representation that was not initially zero-center. Thus, gravity can be included in the zero-center module of osp(4/6); but when this multiplet is restricted to, say, osp(4/4) and reduced, then the submultiplet that contains gravity is not zero-center. It is only by going as high as osp(4/6) in the symmetry hierarchy that gravity becomes zero-center. Furthermore, this feature is lost in osp(4/N) for N > 6, unless we allow for spins exceeding 2.

5. Super Conformal Electrodynamics

The physical states of conformal super electrodynamics form a unitary zero-center module of the superalgebra su(2,2/1).[18] The even part of this superalgebra is su(2,2) ⊗ u(1), and the positive energy representations are accordingly labelled as $D_s(E_o,j_1,j_2/Z_o)$,[42] where (E_o,j_1,j_2) have the same meaning as before and Z_o is the u(1) character. The physical states belong to

$$D_s(\tfrac{3}{2},\tfrac{1}{2},0/-\tfrac{1}{2}) \oplus D_s(\tfrac{3}{2},0,\tfrac{1}{2}/\tfrac{1}{2}) \,. \tag{5.1}$$

The restrictions to the conformal group are

$$D(\tfrac{3}{2},\tfrac{1}{2},0) \oplus D(2,1,0) \,,$$

$$D(\tfrac{3}{2},0,\tfrac{1}{2}) \oplus D(2,0,1) \,. \tag{5.2}$$

The assignments $Z_o = \pm\frac{1}{2}$ are fixed by the requirement that these modules be zero-center modules, as will be seen below. The su(2,2) representations in (5.2) are the massless representations with helicities $j_1 - j_2 = \pm\frac{1}{2}, \pm 1$, associated with neutrinos (or massless leptons) and with photons (or massless vector mesons). We must now construct a field theoretical realization and find the complete Gupta-Bleuler triplets that contain these physical representations of su(2,2/1) as central subquotients.

A very simple realization of su(n,n/1) may be obtained by imbedding it in osp(4n/2). As in Section 3, let (L_{ab}) denote a basis for the symplectic algebra, this time sp(4n) instead of sp(4), so the indices run from 1 to 4n. Operators (κ_{ab}) form some representation of this algebra; it need not be specified for the moment. The Grassmann algebra is generated by $\theta^1,\ldots,\theta^{4n}$. Eqs. (3.1), (3.2), (3.4), (4.1) and (4.2) give us a realization of osp(4n/2) if we replace the term $-2i\theta_a$ by $-2ni\theta_a$ in (3.1) and the term -2 by -2n in (4.2).[31]

Let δ be the indefinite metric that defines su(n,n) in $s\ell(2n,C)$, and take the symplectic metric η to be

$$\eta = \begin{pmatrix} 0 , & \delta \\ -\delta , & 0 \end{pmatrix} . \tag{5.3}$$

Consider the real subalgebra of complexified osp(4n/2) spanned by 2n odd operators $q_a, q_{\dot a}$ and even operators $T_{\dot a b}$ (a,b = 1,...,2n):

$$\begin{aligned} q_a &= \frac{1}{2}(K_a + iQ_a) , \\ q_{\dot a} &= \frac{i}{2}(K_{a+2n} - iQ_{a+2n}) , \\ T_{\dot a b} &= iL_{a+2n,b} - 2\delta_{\dot a b} Z . \end{aligned} \tag{5.4}$$

[The complex structure is defined by the involution $i, q_a, q_{\dot{a}}, T_{\dot{a}b} \to -i, q_{\dot{a}}, q_a, T_{\dot{b}a}$.]

The operators κ_{ab} have not and need not be specified, except insofar as they form a representation of sp(4n,R) in some space V_{in}. The operators (5.4) act on superfields with components in V_{in} and form a realization of su(n,n/1). This action leaves invariant the subspace defined by

$$\partial_{a+2n} \Phi = 0 , \quad a = 1,\ldots,2n , \tag{5.5}$$

and from now on we restrict ourselves to this, much simpler, type of superfield. Then (5.4) reduces to

$$\begin{aligned} q_a &= \partial_a , \quad q_{\dot{a}} = \theta^b(\Omega_{\dot{a}b} - \delta_{\dot{a}b}\theta\cdot\partial) , \\ T_{\dot{a}b} &= \theta_{\dot{a}}\partial_b + \Omega_{\dot{a}b} - \delta_{\dot{a}b}\theta\cdot\partial \quad . \end{aligned} \tag{5.6}$$

Here all indices run from 1 to 2n, $\theta_{\dot{a}} = \delta_{\dot{a}b}\theta^b$, $(\Omega_{\dot{a}b})$ form some representation π_{in} of su(n,n) ⊗ u(1) in a space V_{in}, and the operators (5.6) act on polynomials in $\theta^1,\ldots,\theta^{2n}$ with coefficients in V_{in}. In the usual case of "scalar" superfields (without external indices) on compactified Minkowski space, V_{in} is a space of functions on that space, and $\Omega_{\dot{a}b}$ are the vector fields associated with the action of the conformal group.

Our strategy for constructing zero-center modules is the same as in the orthosymplectic case. If the inducing representation D of su(2,2) ⊗ u(1) is D(1,0,0) ⊗ Id, then the lowest energy carried by the superfield is 0, and the identity representation of su(2,2/1) can be expected to appear as a sub-quotient of the induced representation. In fact, we find that

$$\mathrm{IND}_{su(2,2)\otimes u(1)}^{su(2,2/1)} [D(1,0,0)\otimes Id]$$

$$= D_S(\tfrac{1}{2},\tfrac{1}{2},0/\tfrac{1}{2}) \to \begin{bmatrix} D_S(\tfrac{3}{2},0,\tfrac{1}{2}/\tfrac{1}{2}) \\ \oplus\ Id \\ \oplus\ D_S(\tfrac{3}{2},\tfrac{1}{2},0/-\tfrac{1}{2}) \end{bmatrix} \to D_S(\tfrac{1}{2},0,\tfrac{1}{2}/-\tfrac{1}{2}) \ . \quad (5.7)$$

The reduction on su(2,2) of the physical multiplets was given by (5.2). The reduction of the gauge multiplet on su(2,2)is

$$D(\tfrac{1}{2},0,\tfrac{1}{2}) \oplus D(1,0,0) \oplus D(1,\tfrac{1}{2},\tfrac{1}{2}) \ .$$

6. Extended Super Conformal Electrodynamics

The odd subspace of the superalgebra u(2,2/N) is generated by $(K_a{}^i, K_{\dot a}{}^{*i})$ a = 1,...,4 and i = 1,...,N. The u(2,2) generators $T_{\dot a b}$ (a,b = 1,...,4) and the u(N) generators M^{ij} (i,j = 1,...,N) are restricted in su(2,2/N) by

$$(N-1)\ \delta^{\dot a b}\ T_{\dot a b} = 3M^{ij}\ \delta_{ij} \ . \qquad (6.1)$$

Here δ_{ij} is the Kroenecker symbol and $(\delta_{\dot a b})$ is the su(2,2) indefinite metric. The commutation relations are

$$[K_a{}^i, K_b{}^j]_+ = 0 \qquad ,$$

$$[K_{\dot a}{}^{*i}, K_b{}^j]_+ = T_{\dot a b}\delta^{ij} - \delta_{\dot a b}M^{ij} \quad ,$$

$$[T^{\cdot}_{ab}, K_c^{\ i}] = -\delta^{\cdot}_{ac} K_b^{\ i} + \delta^{\cdot}_{ab} K_c^{\ i} \quad ,$$

$$[T^{\cdot}_{ab}, K^{\cdot}_c{}^{*i}] = \delta^{\cdot}_{cb} K_a^{\ *i} - \delta^{\cdot}_{ab} K^{\cdot}_c{}^{*i} \quad ,$$

$$[M^{ij}, K_a^{\ k}] = -\delta^{ik} K_a^{\ j} + \delta^{ij} K_a^{\ k} \quad ,$$

$$[M^{ij}, K^{\cdot}_a{}^{*k}] = \delta^{jk} K^{\cdot}_a{}^{*i} - \delta^{ij} K^{\cdot}_a{}^{*k} \,. \tag{6.2}$$

There is not yet a simple superfield realization of the superalgebra $\mathfrak{g}$ = su(2,2/N), nor do we have any unitary singletons, so we must discover the zero-center modules by calculations in the enveloping algebra U of $\mathfrak{g}$. In abstract terms, we examine the "K-finite Verma module" generated by the trivial K-type.

Roots and weights are taken to refer to a compact Cartan subalgebra, and the ordering is chosen so that odd and noncompact positive roots increase the energy; that is, the eigenvalue of

$$H = \frac{1}{2} \Sigma\, T^{\cdot}_{aa} \,. \tag{6.3}$$

We examine the space U| > generated by applying U to a "vacuum state" characterized by

$$\mathfrak{g}_- | > = 0 \ , \quad \mathfrak{k} | > = 0 \ , \tag{6.4}$$

where $\mathfrak{g}_-$ is the subalgebra associated with the negative roots and $\mathfrak{k}$ is the compact subalgebra. As a $\mathfrak{g}$-module, U| > can be identified with U($\mathfrak{g}$)/I($\mathfrak{b}$), where I($\mathfrak{b}$) is the ideal generated by the Borel subalgebra $\mathfrak{b} = \mathfrak{g}_- + \mathfrak{k}$.

The energy raising operators (associated with odd positive roots) give the subspace with $E = \frac{1}{2}$ (E = eigenvalue of H), spanned by

$$K_a^{\ i}|\,> = |_a^i> , \quad a = 3,4 ,$$
$$K_a^{*i}|\,> = |_a^i> , \quad a = 1,2 , \qquad (6.5)$$

with i = 1,...,N. These states are ground states; that is, they are annihilated by the energy lowering operators, which proves the existence of extensions by Id of two irreducible representations,

$$\mathrm{Id} \to D_s(\tfrac{1}{2},\tfrac{1}{2},0/\mathrm{spin}+)$$

and

$$\mathrm{Id} \to D_s(\tfrac{1}{2},0,\tfrac{1}{2}/\mathrm{spin}-) ,$$

where spin ± are the two N-dimensional representations of su(N). The first of these, for N = 1, appears as a scalar submodule in (5.7), and the other as gauge submodule. In the helicity conjugate of (5.7) their roles are reversed. It is likely that

$$D_s(\tfrac{1}{2},\tfrac{1}{2},0/\mathrm{spin}+) \ \text{and} \ D_s(\tfrac{1}{2},0,\tfrac{1}{2}/\mathrm{spin}-) \qquad (6.6)$$

are the scalar and gauge modules for all N. These are not unitary.

To describe the higher levels we use the standard labels for irreducible representations of su(N), $w = (n_1,...,n_{N-1})$, where (n_i) is a non-increasing sequence of non-negative integers. Then

$$(\mathrm{spin}+) = (1,0,...) , \quad (\mathrm{spin}-) = (1,1,...,1) .$$

To discover the zero-center modules with $E_o = 1$, we examine the K-finite, zero-center Verma module with minimal weight $(\frac{1}{2},\frac{1}{2},0/\text{spin}+)$. Pushing up with the odd raising operators we find the following two sets of weights:

$$\left.\begin{matrix}(1,1,0)\\(1,0,0)\end{matrix}\right\}\times\left\{\begin{matrix}(2,0,\ldots)\\(1,1,0,\ldots)\end{matrix}\right. \tag{6.7}$$

and

$$(1,\tfrac{1}{2},\tfrac{1}{2})\times\left\{\begin{matrix}(2,1,\ldots,1)\\(0,\ldots)\end{matrix}\right. \tag{6.8}$$

Here su(2,2) weights are on the left and su(N) weights are to the right. All combinations within each set are actually realized. The second order Casimir operator vanishes on the three irreducible representations whose lowest weights are the horizontal combinations in (6.7) and (6.8). We shall now prove that the first two of these are zero-center modules.

Let $|^{i}_{a}\rangle$ and $|^{i}_{a}\rangle$ be the ground states of the K-finite Verma modules with lowest weights $(\frac{1}{2},\frac{1}{2},0/\text{spin}+)$ and $(\frac{1}{2},0,\frac{1}{2}/\text{spin}-)$, respectively, and let the compact subalgebra act on these states precisely as if (7.5) would hold. Then the states associated with the weights (6.7) are

$$K^{*i}_{a}|^{j}_{b}\rangle\,,\quad a,b = 1,2\,. \tag{6.9}$$

The su(2,2) weight is (1,1,0) if we symmetrize in a,b, and (1,0,0) if we antisymmetrize. The su(N) weight becomes (2,0,...) if we symmetrize in i,j, and (1,1,0,...) if we antisymmetrize. The lowering operators K^{*k}_{c}, c = 3,4, obviously annihilate all these states. The lowering operators $K_{c}^{\;k}$, c = 1,2, give

$$(T^{\cdot}_{ac}\delta^{ik} - \delta^{\cdot}_{ac}M^{ik})|^{j}_{b}> = \delta^{ik}\,\delta^{\cdot}_{bc}|^{j}_{a}> - \delta^{jk}\,\delta^{\cdot}_{ac}|^{i}_{b}> .$$

This vanishes if we symmetrize in a,b and in i,j or if we antisymmetrize in both pairs. It follows that nontrivial extensions of $D_s(\frac{1}{2},\frac{1}{2},0/\text{spin}+)$ by either of the following representations exist:

$$D_s(1,1,0\{1\}2,0,...) \text{ and } D_s(1,0,0\{1\}1,1,0,...) . \qquad (6.10)$$

Therefore, both of these are zero-center modules. The first is non-unitary, the second probably unitary.[43] The case N = 1 is clearly exceptional; in that case antisymmetrization in i,j gives zero and there is no unitary zero-center representation with minimal energy $E_o = 1$. The number in $\{\cdot\}$ is the u(1) character.

At the next level, $E_o = \frac{3}{2}$, the only weights that are minimal for unitary representations of su(2,2) are $(\frac{3}{2},\frac{1}{2},0)$ and $(\frac{3}{2},0,\frac{1}{2})$. To get them we push up from (6.10). Let $|^{ij}_{ab}>$ be the ground states of $D_s(1,0,0\{1\}1,1,0,...)$, antisymmetric in i,j and in a,b, with a,b = 1,2. We push up to

$$K^{*k}_{\dot{c}}|^{ij}_{ab}> , \quad c = 1,2 ,$$

$$K_c^{\ k}|^{ij}_{ab}> , \quad c = 3,4 .$$

The first set is obviously annihilated by the lowering operators $K^{*\ell}_{\dot{d}}$, d = 3,4, but the other lowering operators $K_d^{\ \ell}$, d = 1,2 give

$$(T^{\cdot}_{cd}\delta^{k\ell} - \delta^{\cdot}_{cd}M^{k\ell})|^{ij}_{ab}> = -\delta^{\cdot}_{cd}[\delta^{\ell i}|^{kj}_{ab}> - (i,j)] .$$

The second set is annihilated by $K_d^{\ \ell}$, d = 1,2 but $K^{*\ell}_{\dot{d}}$, d = 3,4 give

$$(T_{\dot{d}c}\delta^{k\ell} - \delta_{\dot{d}c}M^{\ell k})|{}^{ij}_{a\dot{b}}> = -\delta_{\dot{d}c}[\delta^{ij}|{}^{\ell j}_{a\dot{b}}> - (i,j)] \, .$$

These vectors have non-zero projections into all symmetry types and we find no extensions. That is, there is no unitary zero-center module with $E_o = \frac{3}{2}$.

The conclusion is that if unitary, zero-center modules exist at all, then they must be

$$D_s(1,0,0\{1\}1,1,0,\ldots,0) \text{ and } D_s(1,0,0\{-1\}1,\ldots,1,0) \, .$$

Conjecture. (i) $D_s(1,0,0\{1\}1,1,0,\ldots)$ and the conjugate $D_s(1,0,0\{1\}1,\ldots,1,0)$ are unitary representations[43] of su(2,2/N) for N > 1, and there is a nondecomposable representation of the form

$$D_s(\tfrac{1}{2},\tfrac{1}{2},0\{\tfrac{1}{2}\}1,0,\ldots) \to \begin{bmatrix} D_S(1,0,0\{1\}1,1,0,\ldots) \\ \oplus \, \mathrm{Id} \\ \oplus \, D_S(1,0,0\{-1\}1,\ldots,1,0) \end{bmatrix} \to D_S(\tfrac{1}{2},0,\tfrac{1}{2}\{-\tfrac{1}{2}\}1,\ldots,1) \, .$$

Acknowledgements

We are very grateful to Gregg Zuckerman for several enlightening discussions; his ideas have provided an important part of the inspiration for this paper. Conversations with Birne Binegar have also been helpful. This work was supported in part by The National Science Foundation.

References.

1. A. D. Sakharov, Dokl. Acad. Nauk SSSR 177, 70 (1967); S. T. Adler, R Mod. Phys. 54, 729 (1982).
2. H. Terazawa, Phys. Rev. D22, 184 (1980); and Refs. 6 and 7.
3. N. T. Evans, J. Math. Phys. 8, 170 (1967); C. Fronsdal, Rev. Mod. Phy 37, 221 (1965).
4. C. Fronsdal, Phys. Rev. D12, 3819 (1975).
5. P. A. M. Dirac, J. Math. Phys. 4, 901 (1963).
6. M. Flato and C. Fronsdal, Lett. Math. Phys. 2, 421 (1978).
7. M. Flato and C. Fronsdal, Phys. Lett. 97B, 236 (1980).
8. G. Mack, Comm. Math. Phys. 55, 1 (1977).
9. G. Mack and I. Todorov, J. Math. Phys. 10, 2078 (1969).
10. B. Binegar, C. Fronsdal, and W. Heidenreich, J. Math. Phys. 24, 2828 (1983).
11. G. Rideau, J. Math. Phys. 19, 1627 (1978).
12. H. Araki, Comm. Math. Phys. 97, 149 (1985).
13. B. Binegar, C. Fronsdal and W. Heidenreich, Ann. Phys. 149, 254 (198
14. E. Angelopoulos, M. Flato, C. Fronsdal, and D. Sternheimer, Phys. Rev D23, 1278 (1981).
15. I. N. Bernshtein, I. M. Gel'fand and S. I. Gel'fand, Funct. Anal. Priozen 5, 1 (1971) [Func. Anal. Appl. 5, 1 (1971)]; G. Pinczon and J. Simon, Rep. Math. Phys. 16, 49 (1979).
16. D. Vogan, "Representations of Real Reductive Lie Groups," Birkhaüser, Boston-Basel-Stuttgart (1981).
17. G. Zuckerman, Ann. Math. 106, 295 (1977).
18. B. Binegar, "On the Unitarity of Conformal Supergravity," UCLA/84/TE preprint, June 1984.
19. M. Kaku, P. K. Townsend and P. van Nieuwenhuizen, Phys. Lett. 69B, (1977).
20. C. Fronsdal, Phys. Rev. D30, 208 (1984).

21. H. Nicolai and E. Sezgin, Nucl. Phys. B242, 69 (1984); L. Castell, W. Heidenreich and T. Künemund, "All Unitary Positive UIRs of osp(N,4)," Starnberg preprint 1984.
22. G. Rideau, Rep. Math. Phys. 16, 251 (1979).
23. R. Akhoury and A. Comtet, Nucl. Phys. B246, 253 (1984).
24. E. Witten, Nucl. Phys. B202, 153 (1982).
25. J. Fang and C. Fronsdal, Phys. Rev. D22, 1361 (1980).
26. Ref. 13, Appendix.
27. J.-P. Gazeau, "Gauge Fixing and Gupta-Bleuler Triplets in de Sitter QED," UCLA/84/TEP/8, July 1984, to appear in J. Math. Phys.
28. P. A. M. Dirac, Ann. Math. 37, 429 (1936).
29. J. Wess and B. Zumino, Nucl. Phys. B70, 39 (1974).
30. B. W. Keck, J. Phys. A8, 1819 (1975).
31. C. Fronsdal, Lett. Math. Phys. 1, 165 (1976).
32. B. Zumino, Nucl. Phys. B127, 189 (1977).
33. C. Fronsdal and T. Hirai, "Unitary Representations of Supergroups," UCLA preprint, April 1985, in this volume.
34. P. Breitenlohner and D. Z. Freedman, Ann. Phys. 144, 249 (1982).
35. D. Z. Freedman and H. Nicolai, Nucl. Phys. B237, 342 (1984).
36. S. Ferrara, J. Wess and B. Zumino, Phys. Lett. B51, 239 (1974).
37. N. Bourbaki, "Groupes et Algebres de Lie" (Hermann, Paris 1975) Chapters 4-8.
38. V. Bargmann, Comm. Pure Appl. Math. 14, 187 (1961), and ibid. 20, 1 (1967).
39. C. Fronsdal, Phys. Rev. D26, 1988 (1982).
40. M. Günaydin and C. Saclioglu, Phys. Lett. 108B, 169 (1982); M. Günaydin, in "Group Theoretical Methods in Physics," Istanbul, 198 (Lecture Notes in Physics 180, Springer-Verlag).
41. H. Nicolai, "Representations of Supersymmetry in Anti-de Sitter Space," CERN TH.3882, April 1984.

42. M. Flato and C. Fronsdal, Lett. Math. Phys. 8, 159 (1984).

43. B. Binegar, "Conformal Superalgebras, Massless Representations and Hidden Symmetries," UCLA/85/TEP/16 preprint, June 1985.

MASSLESS PARTICLES, ORTHOSYMPLECTIC SYMMETRY AND ANOTHER TYPE OF KALUZA-KLEIN THEORY

by

C. Fronsdal

ABSTRACT. The superalgebra osp(8/1) is intimately related to the twistor program. Its most singular representation has the following property: restricted to the conformal subalgebra it contains each and every massless representation exactly once. In other words, one irreducible representation of osp(8/1) describes all massless particles with maximal efficiency. It is believed that such unification is required if massless fields of high spins are to have self-consistent interactions. There are other reasons for studying massless particles of all spins simultaneously. There is a very appealing model in which massless particles are viewed as states of two so(3,2) singletons. The astounding fact is that all free two-singleton states are precisely massless. The most singular representation of osp(8/2) is irreducible on osp(8/1) and completely determined by the latter representation. It finds direct application in supergravity theories. The most interesting Sp(8/R) homogeneous space is 10-dimensional. The action of the conformal subgroup leaves invariant a unique 4-dimensional submanifold that can be identified with space time. Kaluza-Klein expansion of the scalar field on 10-space, around this 4-dimensional manifold, leads to a field theory of massless particles with all integer spins on space time. A supersymmetric extension is also possible.

C. Fronsdal (ed.), Essays on Supersymmetry, 163–265.

0. Introduction

This paper is a study of the geometry and the representations of the orthosymplectic superalgebras osp(2n/1). The reasons for undertaking this investigation are several.

1. The superalgebra osp(8/1) is intimately related to the twistor program. Its most singular representation (the oscillator representation) has the following highly significant property: when it is restricted to the conformal subalgebra (the algebra of infinitesimal conformal transformations of 4-dimensional Minkowski space) then it turns out to contain each and every massless representation of that algebra just once. In other words, one representation of osp(8/1) describes all massless particles with the highest possible efficiency. This representation of osp(8/1) is precisely what one gets by straightforward quantization within the twistor program.

2. Conformally invariant field theories are becoming a strong focus of interest in several domains of physics. The difficulties that one encounters in the quantization of conformal field theories can be ascribed to the singular character of the physically important representations of the conformal group. The simple fact is that the wave operators are never invariant operators, although the free wave equations are invariant. One finds that the typical conformal wave operator is one component of a conformal tensor operator. It should be strongly emphasized that exactly the same phenomenon occurs in N = 2 supersymmetric field theories. These theories are extremely important, since it is believed that they may be finite, but the development of a complete superfield quantum field theory is extraordinarily difficult, for exactly the same reasons that conformal field theories are difficult. The study of conformal field theories within the osp(8/1) context reveals the geometric origin of the problem.

3. There is a good reason for studying massless particles of all spins simultaneously. There is a feeling that massless particles should be regarded as composite objects. In fact, there is a very appealing model, in which massless particles are viewed as states of two so(3,2) singletons. The main point is the astounding fact that all free two-singleton states are precisely massless. In fact, the set of two-singleton states can be precisely identified with the states of the oscillator representation of osp(8/1) discussed above.

4. Recent developments in extended supergravity theories has led to increased interest in massless representations of the orthosymplectic and the superconformal supergroups. The study of osp(2n/1) is of course directly applicable to the orthosymplectic case, for N = 1 and N = 2. In addition, the superconformal algebra su(2,2/1) is a subalgebra of osp(8/2), and in fact all the massless representations of su(2,2/N), for any value of N, can be constructed easily from the oscillator representation of osp(8/2N+1).

Here is a very brief outline of the paper.

Part I. (Sections 1,2). Explains the (well known) construction of the oscillator representation of osp(2n/1) in the context of classical mechanics.

Part II. (Sections 3,4). Sets up the realization of the oscillator representation in terms of superfields.

Part III. (Sections 5-8). Studies highest weight representations of sp(2n,R) and osp(2n/1).

Part IV. (Sections 9-12). Identifies the simplest homogeneous space of sp(2n,R), in two types of descriptions. The first identifies it as the space of Lagrangian planes in 2n-dimensional phase space (twistor space), and the second realizes the same space as the boundary of one of the bounded, symmetric classical domains. Explicit realizations of the oscillator representation on the bundle of half-forms is given, in both descriptions.

Part V. (Sections 13-17). Deals with the physical interpretation, especially in the case n = 4. The homogeneous space is now 10 dimensional, and it can be viewed as a special type of Kaluza-Klein space. It is different from ordinary Kaluza-Klein spaces in that it is "already compact." The action of the conformal group SU(2,2) on this space defines a unique orbit of dimension 4, on which the conformal group acts exactly in the right way for this orbit to be identified with compactified, 4-dimensional Minkowski space. The extra dimensions are dealt with exactly as in Kaluza-Klein theories, but instead of an infinite family of massive particles, one finds a family of massless particles of all spins.

Part VI. (Sections 18-23). Is a study of the much simpler problem of 4-dimensional, scalar conformal field theory. This is normally set up either in the familiar Minkowski notation, or else in the manifestly invariant formulation of Dirac. Here we develop a third description of it, in which spacetime is identified with the U(2) group manifold. This is the distinguished boundary of one of the bounded, classical symmetric domains, and this domain turns out to be precisely the forward tube of relativistic field theories. The Bergman and Cauchy kernels are studied and related to the field theoretical Lagrangian.

Part VII. (Sections 24-26). Returns to sp(8,R) and osp(8/1), and tries to develop the theory in the way that is suggested by Part VI. The paper ends with a discussion of where the next big effort may have to be directed.

I. Geometric Preliminaries

Here we introduce the symplectic algebra sp(2n/R) and the orthosymplectic superalgebra osp(2n/1) in the natural context of Hamiltonian mechanics. This leads directly to the most remarkable representation of osp(2n/1), the oscillator representation. We study the ideal in the enveloping algebra that characterizes this representation.

1. Phase space, sp(2n/R) and osp(2n/1).

Phase space, denoted, V_{2n}, is a real, 2n-dimensional vector space. The natural coordinates

$$q_1,\ldots,q_n;\ p_1,\ldots,p_n = \xi_1,\ldots,\xi_{2n} \tag{1.1}$$

form a basis for the real vector space dual V_{2n}^* that consists of the real linear functions $u = u^a \xi_a$ on V_{2n}. The Poisson bracket

$$\{\xi_a,\xi_b\}_- = -\eta_{ab}\ ;\ (\eta_{ab}) = \begin{pmatrix} 0, & -1 \\ 1, & 0 \end{pmatrix} \tag{1.2}$$

gives rise to the following structures.

(a) A two-form on V_{2n}^* is defined by

$$\langle u,v\rangle = \{u,v\}_- = \eta_{ab}u^a v^b\ ;\ u,v \in V_{2n}^*. \tag{1.3}$$

(b) A bijection $V_{2n} \leftrightarrow V_{2n}^*$ is defined by

$$\langle u,v\rangle = v(\tilde{u}) = -u(\tilde{v})\ ;\ u,v \in V_{2n}^*,\ \tilde{u},\tilde{v} \in V_{2n}. \tag{1.4}$$

This amounts to a rule for raising and lowering indices,

$$\tilde{u}_a = u^b \eta_{ba} , \quad u^a = \eta^{ab} \tilde{u}_b ; \quad \eta^{ab} = \eta_{ab} . \tag{1.5}$$

The tilde will be omitted on the components $\tilde{u}_a$ of $\tilde{u}$.

(c) A family of derivations of $C^\infty(V_{2n})$ is defined by

$$\{f,g\}_- \equiv -\eta_{ab} \frac{\partial f}{\partial \xi_a} \frac{\partial g}{\partial \xi_b} ; \quad f,g \in C^\infty(V_{2n}) . \tag{1.6}$$

To each $f \in C^\infty(V_{2n})$ is thus associated a vector field

$$f^\# = -\eta_{ab} \frac{\partial f}{\partial \xi_a} \frac{\partial}{\partial \xi_b} . \tag{1.7}$$

The space of all $f^\#$, $f \in C^\infty(V_{2n})$ is the Lie algebra of globally Hamiltonian vector fields on V_{2n}.

The Poisson bracket (1.6) turns $C^\infty(V_{2n})$ into a Lie algebra. Our interest focuses on the subalgebra

$$sp(2n) \equiv \{f = c^{ab} \xi_a \xi_b ; \ c^{ab} \text{ real}\} , \tag{1.8}$$

spanned by

$$f_{ab} = \xi_a \xi_b , \quad a,b = 1,\dots,2n . \tag{1.9}$$

The graded Lie algebra osp(2n/1) is the real vector space spanned by f_{ab} and ξ_a, with the following structure

$$\{f_{ab}, f_{cd}\}_- = -\eta_{ac} f_{bd} - \eta_{ad} f_{bc} - \eta_{bc} f_{ad} - \eta_{bd} f_{ac} , \tag{1.10}$$

$$\{f_{ab}, \xi_c\}_- = -\eta_{ac} \xi_b - \eta_{bc} \xi_a , \tag{1.11}$$

$$\{\xi_a, \xi_b\}_+ = 2 f_{ab} . \tag{1.12}$$

The first two brackets are antisymmetric and derived from (1.6). The third is symmetric and defined by (1.12), to be extended by linearity in each argument. The sp(2n) generators f_{ab} span the even subalgebra, and the ξ_a span the odd subspace of osp(2n/1). It is also true that Eq. (1.2) holds, but that relation is not part of the structure of osp(2n/1) as a graded Lie algebra. Nevertheless, it will play a major role in our discussion of the representations of osp(2n/1).

Eq. (1.11) defines an action of sp(2n) in V_{2n},

$$\{f_{ab},\xi_c\}_- = -(\ell_{ab})_c{}^d\,\xi_d\,, \tag{1.13}$$

and by duality an action in V^*_{2n},

$$u \to u\ell_{ab}\,,\quad (u\ell_{ab})^d = u^c\,(\ell_{ab})_c{}^d\,. \tag{1.14}$$

The real matrix algebra spanned by the matrices

$$(\ell_{ab})_c{}^d = \eta_{ac}\delta_b{}^d + \eta_{bc}\delta_a{}^d \tag{1.15}$$

may be identified with sp(2n). If m is a 2n-by-2n real matrix,

$$m = \begin{pmatrix} a & b \\ c & d \end{pmatrix}\,,\quad a,b,c,d \in GL(n,R)\,, \tag{1.16}$$

then the following statements are equivalent:

(i) $m \in sp(2n)$, (ii) $m\eta = {}^t(m\eta)$,

(iii) $\langle um,v\rangle + \langle u,vm\rangle = 0$; $u,v \in V^*_{2n}$,

$$\text{(iv)}\ {}^ta = -d\,,\quad {}^tb = b\,,\quad {}^tc = c\,. \tag{1.17}$$

The compact subalgebra of sp(2n) is isomorphic to u(n). Let h be the matrix

$$h = \frac{1}{4}\sum_a \ell_{aa} = \frac{1}{2}\begin{pmatrix} 0 , & 1 \\ -1 , & 0 \end{pmatrix} \tag{1.18}$$

and let m be as in (1.16); then the following statements are equivalent:

$$\text{(i) } m \in u(n) \quad ,$$

$$\text{(ii) } m \in sp(2n) \cap o(2n) \quad ,$$

$$\text{(iii) } m \in sp(2n) \text{ and } mh = hm \; ,$$

$$\text{(iv) } m \in o(2n) \text{ and } mh = hm \; . \tag{1.19}$$

2. The oscillator representation.

The Heisenberg algebra is the real space of functions on V_{2n} spanned by $\xi_1,...,\xi_{2n}$ and 1, with the structure of Lie algebra given by the Poisson bracket. This algebra has only one irreducible representation (up to equivalence), by self-adjoint operators in a Hilbert space, such that

$$\xi_a \to \hat{\xi}_a \; , \quad 1 \to \text{identity operator} \; .$$

Weyl quantization is an extension of this linear mapping, defined by

$$\xi_a\xi_b \cdots \xi_x \to \text{symm. } \hat{\xi}_a\hat{\xi}_b \cdots \hat{\xi}_x \; ,$$

where complete symmetrization is to be carried out on the indices a,b,...,x.

The oscillator representation of osp(2n/1) is obtained by applying this rule to polynomials of orders 1 and 2:

$$\xi_a \to \hat{\xi}_a\ ,\quad f_{ab} \to \hat{f}_{ab} = \frac{1}{2}(\hat{\xi}_a\hat{\xi}_b + \hat{\xi}_b\hat{\xi}_a) \qquad (= iL_{ab})\ . \tag{2.1}$$

These operators are self-adjoint and form a representation of osp(2n/1), in the sense that Eqs. (1.10)-(1.12) remain valid after the functions ξ_a, f_{ab} are replaced by the operators $\hat{\xi}_a, \hat{f}_{ab}$ and

$$\{\ ,\ \}_- \to \frac{1}{i}[\ .\ ]_-\ ,\quad \{\ ,\ \}_+ \to [\ ,\ ]_+\ , \tag{2.2}$$

where $[\ ,\ ]_-$ is the commutator and $[\ ,\ ]_+$ is the anticommutator.

The operator that corresponds to (1.18),

$$H = \frac{1}{4}\sum_a \hat{f}_{aa}\ , \tag{2.3}$$

is compact and positive; its spectrum is discrete and bounded below. The destruction and creation operators

$$a_j = 2^{-1/2}(\hat{\xi}_j + i\,\hat{\xi}_{n+j})\ ,\quad a_j^* = 2^{-1/2}(\hat{\xi}_j - i\,\hat{\xi}_{n+j})\ , \tag{2.4}$$

$j = 1,...,n$, decrease resp. increase the eigenvalue of H by 1/2. The eigenvector $|0>$ associated with the lowest eigenvalue of H is characterized by

$$a_j|0> = 0\ ;\ \ j = 1,...,n\ . \tag{2.5}$$

It follows from the definitions that

$$H = \frac{1}{2} \Sigma a_j^* a_j + \frac{n}{4}, \tag{2.6}$$

so that the lowest eigenvalue is n/4.

The oscillator representation of osp(2n/1) is realized on the Fock space $\mathcal{H}$ generated from |0> by the operators a_j^*. It is a lowest weight representation and |0> is the vector that is associated with the lowest weight. The maximal compact subalgebra u(n) is represented by operators that commute with H,

$$u(n) = \{\hat{f} = c^{ab} \hat{f}_{ab} ; \hat{f}H = H\hat{f}\} . \tag{2.7}$$

The lowest weight vector |0> is u(n) stable, and each eigenspace of H carries precisely one irreducible representation of u(n).

The restriction of this representation to the even subalgebra is a sum of two irreducible, highest weight representations of sp(2n). In the associated decomposition of Fock space

$$\mathcal{H} = \mathcal{H}_+ + \mathcal{H}_- ,$$

the subspace $\mathcal{H}_+$ (resp. $\mathcal{H}_-$) is generated from |0> by application of an even (resp. odd) number of creation operators. The even subspace $\mathcal{H}_+$ is an irreducible, lowest weight sp(2n) module, with |0> as lowest weight vector. The odd subspace $\mathcal{H}_-$ is also an irreducible, lowest weight sp(2n) module; its lowest weight vector belongs to the space spanned by a_i^*|0>.

This oscillator representation of osp(2n/1) is the only representation that preserves (1.2), in the sense that

$$\frac{1}{i} [\hat{\xi}_a, \hat{\xi}_b] = -\eta_{ab} . \tag{2.8}$$

The operators

$$I_{ab} = [\xi_a, \xi_b] + i\,\eta_{ab} \tag{2.9}$$

thus generate an ideal in the enveloping algebra of osp(2n/1) that uniquely characterizes the oscillator representation by vanishing in it. Therefore, this representation is one of the very singular and very interesting representations that are associated with large ideals in the enveloping algebra. The same is true of the two representations of sp(2n). Indeed, it follows directly from (2.8) that

$$I_{abcd} \equiv (\hat{f}_{ab} - \frac{i}{2}\eta_{ab})(\hat{f}_{cd} - \frac{i}{2}\eta_{cd}) - (\hat{f}_{ac} - \frac{i}{2}\eta_{ac})(\hat{f}_{bd} - \frac{i}{2}\eta_{bd})$$
$$+ i\,\eta_{bc}(\hat{f}_{ad} - \frac{i}{2}\eta_{ad}) \tag{2.10}$$

vanishes in the oscillator representation.

The factor 1/i in (2.2) should be noted. The matrices ℓ_{ab} satisfy exactly the same commutation relations as the differential operators $f^{\#}_{ab}$, and they have exactly the same form as the Poisson bracket relations (1.10). The operators $\hat{f}_{ab}$, on the other hand, satisfy commutation relations with an extra factor of i in them. In general, and especially in the next section, we use symbols iL_{ab} and Q_a for the representatives of f_{ab} and ξ_a in any representation of osp(2n/1). The commutation-anticommutation relations are

$$[L_{ab}, L_{cd}]_- = -(\eta_{ac}L_{bd} + \eta_{ad}L_{bc} + \eta_{bc}L_{ad} + \eta_{bd}L_{ac}) \,, \tag{2.11}$$

$$[L_{ab}, Q_c]_- = -(\eta_{ac}Q_b + \eta_{bc}Q_a) \,, \tag{2.12}$$

$$[Q_a, Q_b]_+ = 2iL_{ab} \,. \tag{2.13}$$

In particular, these are satisfied in the oscillator representation, $iL_{ab} = \hat{f}_{ab}$, $Q_a = \hat{\xi}_a$. The matrices ℓ_{ab} form a representation of sp(2n); that is, $L_{ab} \to \ell_{ab}$ satisfies (2.11).

II. Superfield Preliminaries

Lie algebras can be realized in terms of vector fields on homogeneous spaces. The generalization to superalgebras leads to superfields. Very simple formulas exist for osp(2n/1). The oscillator representation is recovered on an especially degenerate superfield that has no terms beyond the first order in the Grassmann algebra. The connection with induced representations is explained.

3. Superfield representation.

Let X be a homogeneous space for SP(2n). This space may be thought of as space time, to which it will in fact be related later on. Let $\{Q_a, L_{ab}\}$ be a basis for osp(2n/1), so that the oscillator representation is given by $Q_a \to \hat{\xi}_a$ and $iL_{ab} \to \hat{f}_{ab}$, and let κ_{ab} denote the vector field on X associated with L_{ab} via the action of sp(2n) in X. A superfield is a device designed to extend this association to osp(2n/1).

Let θ^a, a = 1,...,2n be the free generators of a Grassmann algebra, so that

$$\theta^a \theta^b + \theta^b \theta^a = 0 .$$

An action of osp(2n/1) in this Grassmann algebra is given by

$$Q_a \to (1 - \frac{i}{2}\,\theta{\cdot}\theta)\,\partial_a + i\theta_a \theta{\cdot}\partial\,, \tag{3.1}$$

$$L_{ab} \to -\,\theta_a\partial_b - \theta_b\partial_a \equiv s_{ab}\,, \tag{3.2}$$

where

$$\partial_a = \partial/\partial\theta^a\,, \qquad \theta_a = \theta^b\,\eta_{ba}\,,$$

$$\theta{\cdot}\theta = \theta^a\,\theta_a\,, \qquad \theta{\cdot}\partial = \theta^a\,\partial_a\,.$$

These operators may be interpreted as "Killing vectors" (and spinors) associated with the "Grassmann metric" η. An equivalent formula for Q_a will be more convenient for generalization, namely

$$Q_a \to \theta^b\,(iL_{ab} + \frac{i}{2}\,\eta_{ab}) + (1 + \frac{i}{2}\,\theta{\cdot}\theta)\,\partial_a\,. \tag{3.3}$$

It is easy to verify that the validity of

$$[Q_a, Q_b]_+ = 2iL_{ab} \tag{3.4}$$

depends only on (3.3) and on the commutation relations between the L_{ab}'s and the θ^a's. The same is true of

$$[Q_a, Q_b]_- + i\eta_{ab} = \theta^c\,\theta^d\,I_{cabd} + 2(1 + \frac{i}{2}\,\theta{\cdot}\theta)^2\,\partial_a\partial_b$$
$$+ 2i(1 + \frac{i}{2}\,\theta{\cdot}\theta)\,\theta^c\,[(L_{ac} + \frac{1}{2}\,\eta_{ac})\,\partial_b - (L_{bc} + \frac{1}{2}\,\eta_{bc})\,\partial_a + \eta_{ab}\partial_c]\,. \tag{3.5}$$

Here I is the tensor defined by Eq. (2.10), with $\hat{f}_{ab}$ replaced by iL_{ab}.

A (scalar) superfield is a mapping of X into the Grassmann algebra; that is, a polynomial in the θ's with coefficients that are sections of fiber bundles over X. Thus, if $x \in X$,

$$\Phi(x) = \phi(x) + \theta^a \psi_a(x) + \frac{1}{2!}\theta^a \theta^b A_{ab}(x) + \frac{1}{3!}\theta^a \theta^b \theta^c B_{abc}(x) \tag{3.6}$$

An action of osp(2n/1) on Φ is given by (3.3), with (3.2) replaced by

$$L_{ab} \to s_{ab} + \kappa_{ab} , \tag{3.7}$$

where κ_{ab} are the vector fields associated with the action of sp(2n) on the homogeneous space X. Here each coefficient is regarded as a scalar field on X. A more satisfactory point of view is to interpret them as the components of multi-spinor fields. In this case L_{ab} is simply the Lie derivative $\mathcal{L}_{ab}$ associated with the vector field κ_{ab},

$$L_{ab}\phi = \kappa_{ab}\phi , \quad L_{ab}(\theta^c \psi_c) = \theta^c (\mathcal{L}_{ab}\psi)_c , \tag{3.8}$$

and so on.

The action of osp(2n/1) on the spinor fields is given by

$$Q_a\Phi = \delta_a\phi + \theta^b \delta_a\psi_b + \frac{1}{2!}\theta^b \theta^c \delta_a A_{bc} + \dots .$$

Using (3.3) we find that

$$\begin{aligned} \delta_a\phi &= \psi_a , \\ \delta_a\psi_b &= i(\kappa_{ab} + \frac{1}{2}\eta_{ab})\phi + A_{ab} , \\ \delta_a A_{bc} &= -i\, M_{abc}(\psi) + B_{abc} , \end{aligned} \tag{3.9}$$

with

$$M_{abc}(\psi) = -(\mathcal{L}_{ab} + \frac{1}{2}\eta_{ab})\,\psi_c + (\mathcal{L}_{ac} + \frac{1}{2}\eta_{ac})\,\psi_b + \eta_{bc}\psi_a$$

$$= -(\kappa_{ab} - \frac{1}{2}\eta_{ab})\,\psi_c + (\kappa_{ac} - \frac{1}{2}\eta_{ac})\,\psi_b - \eta_{bc}\psi_a\ , \tag{3.10}$$

4. Oscillator representation on superfields.

Recall that the oscillator representation is uniquely characterized by the fact that it preserves (1.2); that is,

$$I_{ab} \equiv [Q_a,Q_b]_- + i\eta_{ab} = 0\ . \tag{4.1}$$

To investigate these operators we use Eq. (3.5). Note that if $I_{ab} = 0$, then the operators I_{abcd} must also vanish, as was seen in Section 2. Furthermore, the restriction of the oscillator representation to sp(2n) is a sum of only two irreducible representations, so we expect that the superfield expansion (3.6) may terminate with the second term. The three terms in (3.5) should then vanish separately.

Iteration of (3.9) gives the action of Q_aQ_b:

$$\delta_a\delta_b\phi = i(\mathcal{L}_{ab} - \frac{1}{2}\eta_{ab})\,\phi - A_{ab} \tag{4.2}$$

$$\delta_a\delta_b\psi_c = i(\mathcal{L}_{ab} - \frac{1}{2}\eta_{ab})\,\psi_c - iM_{cab}(\psi) + B_{abc}\ . \tag{4.3}$$

In order to satisfy Eq. (4.1), we have to impose the following restrictions on the superfield:

$$A_{ab} = B_{abc} = \dots = 0\ , \tag{4.4}$$

$$M_{abc}(\psi) = 0\ . \tag{4.5}$$

These conditions are consistent with the action of Q_a on Φ, as (3.9) shows. Eq. (4.5) is equivalent--if (4.4) holds--to the disappearance of the last term in Eq. (3.5).

In particular, (4.5) must hold for spinor fields $\delta_d\psi$ with components

$$\delta_d\psi_c = i(\kappa_{dc} + \frac{1}{2}\eta_{dc})\,\phi\ .$$

Now

$$i\, M_{abc}(\delta_d\psi) = I_{abcd}\,\phi\ ,$$

so it follows that ϕ must carry the oscillator representation. Also, since $\delta_a\phi = \psi_a$ by (3.9), the same is true of the action of (κ_{ab}) on each component of the spinor field.

This result can be rephrased in terms of induced representations. Let π_{in} be the even oscillator representation of sp(2n) in a space V_{in}, and let $\kappa_{ab} = \pi_{in}(L_{ab})$. The operators of the induced representation

$$\pi = \mathrm{IND}\begin{matrix} osp(2n/1) \\ \uparrow \\ sp(2n) \end{matrix}\ \pi_{in}$$

act on superfields with components in V_{in}, with

$$\pi(L_{ab}) = s_{ab} + \kappa_{ab}\qquad ,$$

$$\pi(Q_a) = i\theta^b\,\pi(L_{ab}) + (1 + \frac{i}{2}\,\theta\cdot\theta)\,\partial_a\ .$$

This agrees with (3.7) and (3.3), except for the term $(1/2i)\,\theta_b$ in (3.3).

The agreement becomes complete if we identify the superfield of the induced representation with

$$(1 + \frac{i}{2}\,\theta\cdot\theta)^{-1/2}\ \Phi\ .$$

Therefore, what we have just found, is that this induced representation contains the oscillator representation of osp(2n/1) as an invariant submodule.

III. Algebraic Representation Theory

Various bases are introduced for sp(2n), and important subalgebras are defined. A specific Cartan subalgebra is chosen, and roots and weights are worked out in detail. Lowest weight (positive energy) representations are reviewed, and an exhaustive list of unitarizable ones is given. The important nondecomposable representations associated with the reduction points are described in detail. Similar results are obtained for osp(2n/1), special attention being paid to a family of lowest weight representations that include the oscillator.

5. Lowest weight representations of sp(2n).

Let a_j and a_j^* be defined as in Eq. (2.4),

$$a_j = 2^{-1/2}\,(Q_j + i\,Q_{n+j})\,,\quad a_j^* = 2^{-1/2}\,(Q_j - i\,Q_{n+j})\,, \tag{5.1}$$

and

$$H_{jk} = \frac{1}{2}[a_j, a_k^*]_+ \,, \quad j,k = 1,\dots,n \,, \tag{5.2}$$

$$P_{jk} = \frac{1}{2}[a_j, a_k]_+ \,, \quad P^*_{jk} = \frac{1}{2}[a_j^*, a_k^*]_+ \,, \tag{5.3}$$

In view of (2.13), this is a basis for sp(2n), if we restrict $j \le k$ in (5.3); $\{H_{jj}\}$ is a basis for a compact Cartan subalgebra, (5.2) is a basis for (i times) the (maximal) compact subalgebra, and $\{P^*_{jk}\}$ ($\{P_{jk}\}$) spans the space of positive (negative) root vectors.

The defining representation of sp(2n) is the adjoint action of sp(2n) on the odd subspace of osp(2n/1):

$$[L_{ab}, Q_c]_- = -\eta_{ac}Q_b - \eta_{bc}Q_a \,. \tag{5.4}$$

The weights of this representation are

$$\lambda_j(a_k^*) = \delta_{jk} \,, \quad \lambda_j(a_k) = -\delta_{jk} \,,$$

$$\lambda(a_1^*) > \dots > \lambda(a_n^*) > \lambda(a_n) > \dots > \lambda(a_1) \,.$$

The roots are

$$r_i(P^*_{jk}) = \delta_{ij} + \delta_{ik} \,, \quad r_i(P_{jk}) = -\delta_{ij} - \delta_{ik} \,,$$

$$r_i(H_{jk}) = -\delta_{ij} + \delta_{ik} \,.$$

The positive roots, in descending order,

$$r(P^*_{11}) = (2,0,...,0) \quad ,$$
$$r(P^*_{12}) = (1,1,0,...,0) \quad ,$$
$$...$$
$$r(P^*_{1n}) = (1,0,...,0,1) \quad ,$$
$$r(H_{n1}) = (1,0,...,0,-1) \; ,$$
$$...$$
$$r(H_{21}) = (1,-1,0,...,0) \; ,$$

followed by $\{(0,r')\}$, where r' runs through the positive roots of sp(2n-2). If ρ_n is the half-sum of positive roots of sp(2n), then

$$\rho_n = (n,\rho_{n-1}) = (n,n-1,...,1) \; . \tag{5.5}$$

The weights of finite dimensional representations have integral components. Dominant weights satisfy $\lambda_1 \geq \lambda_2 \geq ... \geq \lambda_n$. The Weyl group consists of all permutations of the components, with any number of sign changes.

The compact subalgebra of sp(2n) is

$$K = u(n) = u(1) \oplus su(n) \; . \tag{5.6}$$

A Cartan subalgebra for su(n) is spanned by the ordered basis ($H_j \equiv H_{jj}$)

$$H_1 - H_n, \; H_2 - H_n, \; ..., \; H_{n-1} - H_n \; , \tag{5.7}$$

and the projection of λ on su(n) is

$$\lambda|su(n) = (\lambda_1 - \lambda_n, \; \lambda_2 - \lambda_n, \; ...) \equiv \bar{\lambda} \; . \tag{5.8}$$

This is a dominant, integral weight for su(n), and λ is said to be K-dominant, if

$$\lambda_j - \lambda_{j+1} = \text{non-negative integer}\ ,\quad j = 1,\ldots,n-1\ . \tag{5.9}$$

Only K-dominant weights will be considered. Eq. (5.9) is equivalent to the statement that the components $\bar{\lambda}_j$ of $\bar{\lambda}$ are non-negative integers, with $\bar{\lambda}_j \geq \bar{\lambda}_{j+1}$, $j = 1,\ldots,n-2$.

The center u(1) of K is generated by

$$H = \frac{1}{2}\sum_{j=1}^{n} H_{jj} = \frac{i}{4}\sum_{a=1}^{2n} L_{aa}\ , \tag{5.10}$$

compare Eq. (2.3). The projection of λ on u(1) is

$$\lambda|u(1) = \frac{1}{2}\sum \lambda_j \equiv E\ ; \tag{5.11}$$

this number will be called the energy of λ. There is no restriction of integrality on E.

A K-finite sp(2n) module is a representation of sp(2n), with the property that its restriction to K is a countable direct sum of finite dimensional representations. From now on this property will always be taken for granted. An sp(2n) module is called a lowest weight representation, with K-dominant lowest weight λ, if it contains a cyclic vector $|\lambda\rangle$, such that

$$H_{jj}|\lambda\rangle = \lambda_j|\lambda\rangle\ ,\quad P_{jk}|\lambda\rangle = 0\ , \tag{5.12}$$

for j,k = 1,...,n. An irreducible lowest weight representation is fixed up to equivalence by the lowest weight.

Let $\mathcal{D}(\lambda)$ be the irreducible, lowest weight representation of sp(2n) with K-dominant lowest weight λ. We need to know the range of values of λ for which this is the differential of a unitary representation of a covering of SP(2n). When it is we say that $\mathcal{D}(\lambda)$ is unitarizable. Define $\bar{\lambda}$ and E as in (5.8) and (5.11), fix $\bar{\lambda}$, and consider the family $\{\mathcal{D}(\lambda)\}$ indexed by E real. It is well known that $\mathcal{D}(\lambda)$ is unitarizable when E is sufficiently positive. The complete answer is found in recent work of Enright, Howe and Wallach.

Let q,r be natural numbers

$$n \geq r \geq q \geq 1 ,$$

defined by

$$\bar{\lambda} = (\ldots,1,\ldots,1,0,\ldots,0) , \tag{5.13}$$

in which q-1 zeros are preceded by precisely r-q ones. Also, let

$$z = n - \lambda_n .$$

[The notation is basically that of Enright, Howe and Wallach, except that we had to replace $\lambda_k \to -\lambda_{n+1-k}$ to compensate for the fact that they dealt with highest weight representations.]

<u>Theorem.</u> $\mathcal{D}(\lambda)$ is unitarizable if and only if z belongs to the set that consists of

$$\text{(i)} \quad z \leq \frac{1}{2}(r+1) \quad ,$$

$$\text{(ii)} \quad 2z - r - 1 = 1,2,\ldots,q-1 . \tag{5.14}$$

The corresponding range of E is found from

$$2E = \Sigma\,\bar{\lambda}_j + n(n-z)\ .$$

From Section 2 we obtain the lowest weights of the two oscillators:

$$2\lambda = (1,1,...,1)\ ,\quad \bar{\lambda} = (0,0,...)\ ,$$

$$2\lambda = (3,1,...,1)\ ,\quad \bar{\lambda} = (1,0,...)\ ,$$

for the even and odd oscillator, respectively. "Off-shell" continuation leads to the families

$$2\lambda = (\alpha,\alpha,...,\alpha)\quad ,\quad q = r = n\quad ,$$

$$2\lambda = (\alpha+2,\alpha,...,\alpha)\ ,\quad q+1 = r = n\ .$$

Among these, the unitary representations are those for which one or the other of the following conditions are satisfied:

(i) $\alpha \geq n-1$ (discrete holomorphic series) ,

(ii) $\alpha = 1,2,...,n-2$.

In the even case we must add the point $\alpha = 0$, the trivial representation. The corresponding values of E are given by

$$\alpha = 4E/n\ ,\quad \text{even case}\quad ,$$

$$\alpha = 4(E - \tfrac{1}{2})/n\ ,\ \text{odd case}\ .$$

When n = 2 the unitary range for the oscillator families is

$$E \geq \tfrac{1}{2} \text{ and } E = 0 \text{ , even case ,}$$

$$E \geq 1 \qquad\qquad \text{, odd case .}$$

When n = 4

$$E \geq 3 \text{ and } E = 0,1,2 \text{ , even case ,}$$

$$E \geq \tfrac{7}{2} \text{ and } E = \tfrac{3}{2}, \tfrac{5}{2} \text{ , odd case .}$$

6. K-structure.

The K-structure of a lowest weight representation $\mathcal{D}(\lambda)$ of sp(2n) is its reduction on the compact subalgebra K = u(n) = u(1) ⊕ su(n),

$$\mathcal{D}(\lambda)|u(n) = \sum_{\mu} d(\mu) = \sum_{E,\overline{\mu}} d(E,\overline{\mu}) \text{ ,}$$

where $d(\mu) = d(E,\overline{\mu})$ is an irreducible representation of u(n), with u(1) character E and su(n) highest weight $\overline{\mu}$. More precisely, if E_o is the energy of λ,

$$\mathcal{D}(\lambda)|u(n) = \sum_{k=0,1,\ldots} \sum_{\overline{\mu}\in\Lambda(k)} d(E_o+k,\overline{\mu}) \text{ .} \tag{6.1}$$

The set $\Lambda(k)$ of dominant integral su(n) weights may include repetitions. We shall calculate $\Lambda(k)$ explicitly in the simplest cases.

Let d_a^* be the defining representation of u(n), with energy +1/2 and su(n) highest weight (1,0,...,0) on the basis (5.7). This is the representation that is determined by the adjoint action of u(n) on a_j^*:

$$[H_{jk}, a_i^*] = \delta_{ji} a_k^* \ .$$

The oscillator representation is realized on the Fock space that is constructed by the free, commutative action of the a_j^* on the lowest weight vector (the Fock vacuum), therefore the K-structures of the even and odd parts are

$$\mathcal{D}(\tfrac{1}{2},\tfrac{1}{2},\ldots,\tfrac{1}{2})|u(n) = \sum_k (d_a^* \otimes)_S^{2k} \otimes d(n/4,0) \quad ,$$
$$\mathcal{D}(\tfrac{3}{2},\tfrac{1}{2},\ldots,\tfrac{1}{2})|u(n) = \sum_k (d_a^* \otimes)_S^{2k+1} \otimes d(n/4,0) \ . \tag{6.2}$$

The subscript S means that only the completely symmetric part of the direct product is to be retained. Now these summands are irreducible, with su(n) highest weights (2k,0,...,0) and (2k+1,0,...,0). Therefore, for the oscillators, the structure of $\Lambda(k)$ is particularly simple, containing only one weight:

$$\text{Even osc.:} \quad \Lambda(k) = \{\bar{\mu} = (2k,0,\ldots,0)\} \quad ,$$

$$\text{Odd osc.:} \quad \Lambda(k) = \{\bar{\mu} = (2k+1,0,\ldots,0)\} \ . \tag{6.3}$$

In the general case we have to deal with another Fock space. Let B be the Borel subalgebra of sp(2n) that is spanned by u(n) and the negative root vectors P_{jk}. Let $d(\lambda)$ be an irreducible, finite dimensional u(n) module with highest weight λ and highest weight vector $|\lambda\rangle$. Promote this u(n) module to a B module by setting $P_{jk} \to 0$. Now let $\mathcal{F}(\lambda)$ be the space generated by the action of the positive root vectors P_{jk}^* on $d(\lambda)$. As an sp(2n) module it is

$$\mathcal{F}(\lambda) = U \underset{B}{\otimes} d(\lambda) , \tag{6.4}$$

where U is the enveloping algebra of sp(2n). The K-structure of $\mathcal{F}(\lambda)$ is

$$\mathcal{F}(\lambda)|u(n) = \sum_k (d_P{}^*\otimes)_S{}^k \otimes d(\lambda) , \tag{6.5}$$

where $d_P{}^*$ is the representation of u(n) that is defined by the adjoint action of u(n) on the P^*_{jk},

$$d_P{}^* = (d_a{}^*\otimes)_S{}^2 .$$

<u>Definition.</u> Consider the family $\{\mathcal{F}(\lambda) = \mathcal{F}(E_o,\bar{\lambda})\}$, with $\bar{\lambda}$ fixed, indexed by E_o real. We say that $\mathcal{F}(\lambda)$, or λ, or E_o, is at a reduction point, if $\mathcal{F}(\lambda)$ is reducible. The first reduction point is the highest value of E_o for which $\mathcal{F}(\lambda)$ is reducible.

<u>Theorem.</u> Within the range of unitary representations, more precisely, for $z \leq (r+q)/2$ in the terminology of Section 5, the only reduction points are at the end of the continuous range and at the isolated unitary points given by (5.14).

Consequently, except at these exceptional points, $\mathcal{F}(\lambda)$ is irreducible and equal to $\mathcal{D}(\lambda)$, and (6.5) gives the K-structure of the latter. This structure is quite complicated, even for the oscillator families; it simplifies at the first reduction point and progressively at the higher reduction points.

At a reduction point $\mathcal{F}(\lambda)$ becomes reducible, but never decomposable. Therefore, at any reduction point, $\mathcal{F}(\lambda)$ contains an invariant submodule $\mathcal{F}_o(\lambda')$ and

$$\mathcal{F}(\lambda) = \mathcal{D}(\lambda) \rightarrow \mathcal{F}_0(\lambda') \,. \tag{6.6}$$

[This is our notation for non-decomposable representations. The whole is an extension of the first by the second, the arrow indicates the direction of the leak into the invariant subspace.] The submodule has a lowest weight λ' and may be reducible or irreducible. The relative simplicity of $\mathcal{D}(\lambda)$ at reduction points is due to the secession of $\mathcal{F}_0(\lambda')$. Our interest focuses on the evaluation of λ' at each (unitary) reduction point.

7. Non-decomposable representations of sp(2n).

Eq. (6.6) implies that $\mathcal{D}(\lambda)$ and every component of $\mathcal{F}_0(\lambda')$ have the same infinitesimal character; that is, the same values of the Casimir operators. In particular, this must be true of $\mathcal{D}(\lambda)$ and $\mathcal{D}(\lambda')$, which means that λ and λ' are Weyl equivalent.

Definition. Let ρ_n be the half-sum of positive roots given by (5.5), and $\tilde{\rho}_n$ the conjugate,

$$\tilde{\rho}_n = (1,2,...,n) \,. \tag{7.1}$$

Two weights λ and λ' are said to be Weyl equivalent if $\lambda - \tilde{\rho}_n$ and $\lambda' - \tilde{\rho}_n$ are on the same orbit of the Weyl group. In that case we write $\lambda \simeq \lambda'$.

Applyingmethodsadaptedfromthepaperby Enright, Howeand Wallach, we easily prove the following. The parameter q is as in Section 5.

Theorem. There is a unique sequence $r^1,...,r^q$ of positive, noncompact roots, such that, at the p'th reduction point, the lowest weights of $\mathcal{F}(\lambda)$

and $\mathcal{F}_o(\lambda')$ are related by

$$\lambda' = \lambda + r^1 + r^2 + \dots + r^{q-p+1} .$$

We shall calculate $r^1, \dots, r^q$ for the two oscillator families. In the even case, at the pth reduction point,

$$2\lambda = (n-p, n-p, \dots) \quad ,$$

$$2\lambda - 2\tilde{\rho} = (n-p-2, n-p-4, \dots) .$$

At the last reduction point, $p = q = n$, $2\lambda - 2\tilde{\rho} = (-2,-4,\dots)$. According to the theorem,

$$\sum_j (2\lambda' - 2\lambda)_j = \sum_j 2r_j^1 = 4 .$$

Recall that the Weyl group consists of permutations and any number of sign changes. The only sign change that, when applied to $(-2,-4,\dots)$ increases the sum of the components by 4, is the reversal of the first component. Hence $r^1 = (2,0,\dots) = r(P_{11}^*)$. In the same way one finds

$$r^j = r(P_{jj}^*) , \quad j = 1,\dots,q = n .$$

The pair λ,λ' at the p'th reduction point is

$$2\lambda = (n-p, n-p , \dots) \quad ,$$

$$2\lambda' = (n-p+4, \dots, n-p+4, n-p, \dots, n-p) ,$$

the difference extending to the (n+1-p) 'th place.

When n = 2 the two pairs are

$$\lambda = (\tfrac{1}{2},\tfrac{1}{2}) \ , \quad \lambda' = (\tfrac{5}{2},\tfrac{5}{2}) \ ;$$

$$\lambda = (0,0) \ , \quad \lambda' = (2,0) \ .$$

When n = 4 the four pairs are

$$\lambda = (\tfrac{3}{2},\tfrac{3}{2},...) \ , \quad \lambda' = (\tfrac{7}{2},\tfrac{7}{2},\tfrac{7}{2},\tfrac{7}{2}) \ ;$$

$$\lambda = (1,1,...) \ , \quad \lambda' = (3,3,3,1) \ ;$$

$$\lambda = (\tfrac{1}{2},\tfrac{1}{2},...) \ , \quad \lambda' = (\tfrac{5}{2},\tfrac{5}{2},\tfrac{1}{2},\tfrac{1}{2}) \ ;$$

$$\lambda = (0,0,...) \ , \quad \lambda' = (2,0,0,0) \ .$$

In the case of the odd oscillator representation, at the p 'th reduction point,

$$(2\lambda - 2\tilde{\rho}) = (n\text{-}p,\ n\text{-}p\text{-}4,\ n\text{-}p\text{-}6,\ ...) \ .$$

At the last reduction point, p = q = n-1, this becomes (1,-3,-5,...). In the same way as in the even case one observes that $r^1 = (1,1,0,...)$ and

$$r^j = r(P^*_{jj+1}) \ , \qquad j = 1,...,q = n\text{-}1 \qquad ,$$

$$2\lambda = (n\text{-}p\text{+}2,\ n\text{-}p,\ n\text{-}p,\ ...) \qquad ,$$

$$2\lambda' = (n\text{-}p\text{+}4,\ n\text{-}p\text{+}4,\ ...,\ n\text{-}p\text{+}4,\ n\text{-}p\text{+}2,\ n\text{-}p,\ ...) \ ,$$

the difference extending to the (n+1-p) 'th place.

When n = 2 the only reduction point is

$$\lambda = (\tfrac{3}{2},\tfrac{1}{2}) \qquad \lambda' = (\tfrac{5}{2},\tfrac{3}{2}) .$$

When n = 4 the 3 pairs are

$$\lambda = (\tfrac{3}{2},\tfrac{1}{2},\tfrac{1}{2},\tfrac{1}{2}) \qquad \lambda' = (\tfrac{5}{2},\tfrac{5}{2},\tfrac{5}{2},\tfrac{3}{2}) \quad ;$$

$$\lambda = (1,0,0,0) \qquad \lambda' = (2,2,1,0) \quad ;$$

$$\lambda = (\tfrac{1}{2},-\tfrac{1}{2},-\tfrac{1}{2},-\tfrac{1}{2}) \qquad \lambda' = (\tfrac{3}{2},\tfrac{1}{2},-\tfrac{1}{2},-\tfrac{1}{2}) .$$

8. Lowest weight representations of osp(2n/1).

The lowest weight representations of osp(2n/1) are in a simple and one-one correspondence with those of sp(2n). There are no unitarizable highest weight representations.

The Cartan subalgebra of sp(2n) is also the Cartan subalgebra of osp(2n/1). The roots of osp(2n/1) include, besides the roots of sp(2n) (henceforth called even roots), the following odd roots:

$$r_j(a_k^*) = \lambda_j(a_k^*) = \delta_{jk} , \quad r_j(a_k) = \lambda_j(a_k) = -\delta_{jk} .$$

The half-sum of negative odd roots is (-1/2,-1/2,...), and addition of this to ρ_n, Eq. (5.5), gives the quantity that takes the place of ρ_n in the character formula for osp(2n/1), namely

$$\rho_n^{\,s} = (n - \tfrac{1}{2}, n - \tfrac{3}{2}, \ldots, \tfrac{1}{2}) . \tag{8.1}$$

Let $\mathcal{D}^S(\lambda)$ be an irreducible lowest weight representation of osp(2n/1) with K-dominant lowest weight λ. Then $\mathcal{D}^S(\lambda)$ is fixed up to equivalence by λ. There is a unique, cyclic vector $|\lambda>$ such that (5.12) holds, and in addition

$$a_j|\lambda> = 0 , \quad j = 1,...,n .$$

The restriction of $\mathcal{D}^S(\lambda)$ to sp(2n) is a finite sum of irreducible, lowest weight sp(2n) modules, among which the lowest of the lowest weights is λ, this appears in the reduction with multiplicity 1.

Unitarizability of any representation $\mathcal{D}(\lambda)$ of sp(2n), that belongs to the discrete part of (5.14), can be proved most easily by showing that it occurs in the discrete reduction of a tensor product of oscillator representations. The same method can be applied to show the unitarizability of some of the representations $\mathcal{D}^S(\lambda)$ of osp(2n/1). Unitarizability of $\mathcal{D}(\lambda)$ is a necessary condition, but it is not sufficient. If $\mathcal{D}(\lambda)$ is the odd oscillator representation, then $\mathcal{D}^S(\lambda)$ is not unitarizable.

As in Section 6 for sp(2n), we here define the K-finite lowest weight representation

$$\mathcal{F}^S(\lambda) = U^S \otimes_{B^S} d(\lambda) ,$$

of osp(2n/1). Here U^S is the enveloping algebra of osp(2n/1), B^S is the Borel sub-superalgebra generated by B and the negative root vectors a_j, and a_j acts trivially in $d(\lambda)$. The restriction of $\mathcal{F}^S(\lambda)$ to sp(2n) is a finite sum:

$$\mathcal{F}^S(\lambda)|sp(2n) = \sum_{\lambda' \in \Xi(\lambda)} \mathcal{F}(\lambda') .$$

The set $\Xi(\lambda)$ of K-dominant lowest weights is easily determined. Let S_1^+ be the Grassmann algebra (with unit) generated by $\{a_j^*\}$ $j = 1,...,n$, with the action of u(n) that is induced by the adjoint action on the linear part. Then we have the following identity of u(n)-modules:

$$\sum_{\lambda' \in \Xi(\lambda)} d(\lambda') = S_1^+ \otimes d(\lambda) .$$

As in Section 7, consider the family $\{\mathcal{F}^S(\lambda)\}$ with $\bar{\lambda}$ fixed and E_o real. It will turn out that $\mathcal{F}^S(\lambda)$ is irreducible except for some isolated, small values of E_o, the reduction points. At such a point

$$\mathcal{F}^S(\lambda) = \mathcal{D}^S(\lambda) \to \mathcal{F}_o{}^S(\lambda') ,$$

where $\mathcal{F}_o{}^S(\lambda')$ is a maximal ideal with lowest weight $\lambda' > \lambda$. The existence of this extension of $\mathcal{D}^S(\lambda)$ by $\mathcal{F}_o{}^S(\lambda')$ implies that $\mathcal{D}^S(\lambda)$ and $\mathcal{D}^S(\lambda')$ have the same infinitesimal character, which is to say that they must be Weyl equivalent, or more precisely super Weyl equivalent.

<u>Definition</u>. Let $\tilde{\rho}_n{}^S$ be the conjugate of (8.1):

$$2\tilde{\rho}_n{}^S = (1,3,...,2n-1) .$$

Two weights λ and λ' are said to be super Weyl equivalent if $\lambda - \tilde{\rho}_n{}^S$ and $\lambda' - \tilde{\rho}_n{}^S$ are on the same Weyl orbit. [The Weyl group is that of sp(2n).]

We don't know all the reduction points, but we do know those at which $\mathcal{D}^S(\lambda)$ is unitarizable. We limit the discussion to the isolated, unitary points. In the case of the oscillator family they are

$$2\lambda = (n\text{-}p,\ n\text{-}p,\ \ldots) \quad ,$$

$$2\lambda - 2\tilde{\rho}^S = (n\text{-}p\text{-}1,\ n\text{-}p\text{-}3,\ \ldots)\ .$$

At the last point, $p = n$, $\lambda = 0$ and since $a_j a_k^*|0\rangle = 0$, $\lambda' - \lambda = r(a_1^*)$. Generally,

$$\lambda' = \lambda + r(a_1^*) + \ldots + r(a^*_{n-p+1})\ .$$

Furthermore, for the oscillator family at least, these are the only unitary reduction points. For $n = 2$ the two pairs are

$$\lambda = (\tfrac{1}{2},\tfrac{1}{2}) \ , \quad \lambda' = (\tfrac{3}{2},\tfrac{3}{2}) \quad ,$$

$$\lambda = (0,0) \ , \quad \lambda' = (1,0) \quad .$$

For $n = 4$ the four pairs are

$$\lambda = (\tfrac{3}{2},\tfrac{3}{2},\ldots) \ , \quad \lambda' = (\tfrac{5}{2},\tfrac{5}{2},\tfrac{5}{2},\tfrac{5}{2}) \quad ;$$

$$\lambda = (1,1,\ldots) \ , \quad \lambda' = (2,2,2,1) \quad ;$$

$$\lambda = (\tfrac{1}{2},\tfrac{1}{2},\ldots) \ , \quad \lambda' = (\tfrac{3}{2},\tfrac{3}{2},\tfrac{1}{2},\tfrac{1}{2}) \quad ;$$

$$\lambda = 0 \qquad , \quad \lambda' = (1,0,0,0) \quad .$$

IV. Homogeneous Space and Line Bundle

The simplest and most important sp(2n) homogeneous space is the space of maximal isotropic subspaces of phase space. The dimension is n(n+1)/2; 10 in the most important case n = 4. A well known parameterization in terms of symmetric, unitary matrices is described, and the action of SP(2n,R) is worked out in complete detail. The oscillator representation appears on a line bundle of half-forms over this manifold.

9. The homogeneous space X.

It is necessary to make a definite choice of the SP(2n) homogeneous space X. The selection is difficult because the criteria are vague. The physical requirement that X must be interpretable in terms of ordinary space time suggests the lowest possible dimensionality. Our choice is mainly based on a criterion of simplicity.

Let V be a finite dimensional, irreducible SP(2n) module, z an element of V and Z the SP(2n) orbit that contains z. Then Z is an SP(2n) homogeneous space. Suppose that the oscillator representation $D(\frac{1}{2},...,\frac{1}{2})$ can be realized by operators acting in a Hilbert space that consists of sections of a line bundle over Z. Let f denote a section associated with the lowest weight, and let ϕ be a homogeneous function on V such that (locally) the restriction of ϕ to Z is f. Now f obeys (j,k=1,...,n)

$$P_{jk}f = 0\,, \quad H_{jk}f = \frac{1}{2}\, f\, \delta_{jk}\,,$$

and therefore ϕ satisfies

$$P_{jk}\phi^{-2} = 0\,,\quad H_{jk}\phi^{-2} = -\phi^{-2}\,\delta_{jk}\,.$$

This makes ϕ^{-2} the lowest weight vector of a finite dimensional representation equivalent to $\mathcal{D}(-1,\ldots,-1)$. The simplest possibility is that ϕ^{-2} is linear, $\phi^{-2} \in V^*$, then as an sp(2n) module,

$$V^* = \mathcal{D}(-1,\ldots,-1)\,. \tag{9.1}$$

What this means, in simple if imprecise terms, is that we may expect to find the oscillator representation realized on a space of functions, homogeneous of degree -1/2, on the space Z.

We shall assume, with small loss in generality and great gain in simplicity, that our homogeneous space X is associated with an orbit of SP(2n,R) in V. The lowest weight of $\mathcal{D}(-1,-1,\ldots)$ is the lowest weight of the n'th exterior power of the defining representation of sp(2n); hence V is the space of exterior n-forms on phase space.

Let V^ℓ be the subset of V that consists of exterior n-forms of rank ℓ. The action of SP(2n) in V leaves each V^ℓ invariant. Among these manifolds, the one with lowest dimensionality is V^n, the space of n-forms of minimal rank. We therefore restrict our attention to SP(2n) orbits in V^n.

Every z in V^n has a factorization in terms of n elements of V^*_{2n}:

$$z = v(1) \wedge \ldots \wedge v(n)\,,\quad v(j) \in V^*_{2n}. \tag{9.2}$$

The v(j) are not determined by z, but the n-plane spanned by them is. Hence we have a mapping β:

$$z \to \beta(z) = \mathrm{Span}\{\tilde{v}(1),\ldots,\tilde{v}(n)\} \tag{9.3}$$

of V^n onto the space of n-planes through the origin of V_{2n}. The vectors $\tilde{v}(j)$ in V_{2n} are related to the co-vectors v(j) as in Eq. (1.5).

We can define a two-form on V_{2n} by means of (1.3) and (1.4): $<\tilde{u},\tilde{v}> = <v,u>$; it is preserved by the action of SP(2n) in V_{2n}. We say that $\beta(z)$ is isotropic, that z is isotropic, and that the SP(2n) orbit through z is isotropic, if this form vanishes on $\beta(z)$.

Definition. Let Z be the space of isotropic n-forms of rank n on phase-space. From now on, $X = \beta(Z)$, the space of maximal isotropic subspaces of V_{2n}; that is, the space of Lagrangian subspaces of V_{2n}.

The identification of X as an SP(2n) homogeneous space and as the distinguished boundary of a classical symmetric domain will be discussed later.

Before continuing our study of the general case, we pause to consider the physical meaning of X in the simplest case. The conformal group of $M_{2,1}$, three-dimensional Minkowski space-time, is locally isomorphic to SO(3,2) and to SP(4). When n = 2, Z is the space of isotropic 2-forms of rank 2. If a two-form z has components z^{ab}, a,b, = 1,...,4, then it is isotropic if and only if

$$\eta_{bc} z^{ab} z^{cd} = 0 , \tag{9.4}$$

in which case it is either of rank 2 or else identically zero. Let (γ_α), α = 0,1,2,3,5, be a set of real 4-by-4 Dirac matrices, satisfying

$$\begin{aligned} &[\gamma_\alpha,\gamma_\beta]_+ = -2\delta_{\alpha\beta} , \quad \delta = \text{diag.}(+1,-1,-1,-1,+1) , \\ &(\gamma_\alpha\eta)_{ab} = -(\gamma_\alpha\eta)_{ba} . \end{aligned} \tag{9.5}$$

We map Z bijectively to the 3+2 -cone:

$$y_\alpha = (\gamma_\alpha \eta)_{ab} z^{ab} \quad , \tag{9.6}$$

$$y^2 \equiv y_0^2 - y_1^2 - y_2^2 - y_3^2 + y_5^2 = 0 \, . \tag{9.7}$$

The action of SP(4) on (y_α) is the defining representation of SO(3,2). The space Z can therefore be identified with this 5-cone; it is the analogue, for 3-dimensional space time, of Dirac's conformal 6-cone. The manifold X is the projective cone, obtained by identifying λy with y for all real $\lambda \neq 0$. An alternative projection, defined in the same way but with $\lambda > 0$, leads to the space $S^1 \times S^2$ that is parameterized by an angle in the 0,5 plane and a unit vector in Euclidean 3-space. The projective cone X is therefore $S^1 \times S^2 / Z_2$; it is the conformal compactification of 3-dimensional Minkowski space time.

The intimate relationship between the following statements about 3-dimensional, massless physics is noteworthy. (a) The space X contains no extra dimensions. (b) The components ϕ, ψ of the superfield are fields on space time. (c) The oscillator representation of osp(4/1), restricted to the 3-dimensional Poincaré group, has only two irreducible components. (d) The 3-dimensional Poincaré group has only two massless representations. (e) Massless physics in 3-dimensional space is easy.

The projective cone X can be identified with the space of 2-by-2 symmetric, unitary matrices. An explicit map is

$$w = (y_5 + iy_0)^{-1} \, (y_3 + i\sigma_1 y_1 + i\sigma_3 y_2) \, . \tag{9.8}$$

This is the most convenient type of parameterization to use in the general case, if n > 2. It will be discussed next.

10. Parameterization of X.

In this section, $\xi, \xi', \ldots$ are elements of V_{2n}, with components

$$\xi_1, \ldots, \xi_{2n} = q_1, \ldots, q_n \; ; \; p_1, \ldots, p_n \; .$$

We think of ξ as a column vector in two ways, as an element of the real vector space V_{2n}:

$$\xi = \begin{bmatrix} \xi_1 \\ \vdots \\ \xi_{2n} \end{bmatrix} \in V_{2n} \, ,$$

or as an element of a complex, n-dimensional vector space W_n:

$$\xi = \begin{bmatrix} q_1 \\ \vdots \\ q_n \end{bmatrix} + i \begin{bmatrix} p_1 \\ \vdots \\ p_n \end{bmatrix} = q + ip \in W_n \, .$$

Thus ξ is real if $p = 0$, and $\xi^* = q - ip$. The space X contains just one real element, the n-plane

$$x_o = \{\xi \in W_n ; \, \mathrm{Im}\, \xi = 0\} \; ; \tag{10.1}$$

this will serve as a fixed reference point in X.

The action of SP(2n) in V_{2n} is given by

$$\xi \to g\xi \, , \quad g = \begin{pmatrix} A & B \\ C & D \end{pmatrix} , \tag{10.2}$$

where A,B,C,D are n-by-n real matrices. We have

$$g \in SP(2n) \Longleftrightarrow {}^t g \eta g = \eta \,, \quad \eta = \begin{pmatrix} 0 & -1 \\ 1 & 0 \end{pmatrix} ,$$

$$\therefore {}^t CA \text{ and } {}^t DB \text{ symmetric} \; ; \; {}^t DA - {}^t BC = 1 \,. \qquad (10.3)$$

Note the following subgroups of SP(2n):

$$g \in U(n) \Longleftrightarrow g \in SP(2n) \,, \quad {}^t gg = 1 \,; \qquad (10.4)$$

$$g \in O(n) \Longleftrightarrow g \in U(n) \quad , \quad B = C = 0 \,. \qquad (10.5)$$

If $g \in U(n)$, and $\xi \to g\xi$ is interpreted as a transformation in W_n, then it is unitary. If $g \in O(n)$, then it is real orthogonal. The action of SP(2n) in X is given by

$$x \to gx = \{g\xi ; \xi \in x\} \,. \qquad (10.6)$$

Now U(n) acts transitively in X. In fact, let F_o be an orthonormal basis in x_o; then F_o is an orthonormal basis for W_n. Let x be an element of X, and choose an orthonormal basis F for W_n, consisting of vectors belonging to x. This is possible, since the dimension of x is n. Then there is a unitary transformation that takes F_o to F, and hence an element of U(n) that takes x_o to x. The stabilizer of x_o in U(n) is O(n), so

$$X = U(n)/O(n) \,. \qquad (10.7)$$

In particular, dim X = n(n+1)/2.

The kernel of the mapping

$$U(n) \ni g \to w = g^* g^{-1} \in U(n)$$

is O(n); so the image can be identified with X:

$$X = \{w = g^* g^{-1} \ ; \ g \in U(n)\}$$

$$= \{w \in U_n \quad ; \ {}^t w = w\} \quad . \tag{10.8}$$

Indeed, let $x = gx_0$, $g \in U(n)$; then ξ belongs to x if and only if $\alpha^{-1}\xi$ is real, $\xi^* = g^* g^{-1}\xi = w\xi$. This gives us a parameterization of X; we shall now calculate the action of SP(2n) on w.

If $\xi^* = w\xi$, then Im $\xi = \sigma$ Re ξ, where σ,

$$\sigma = i(w-1)/(w+1) \ , \quad w = (i+\sigma)/(i-\sigma) \tag{10.9}$$

is a real, symmetric matrix. We write

$$\xi = \begin{pmatrix} q \\ \sigma q \end{pmatrix}$$

and evaluate

$$g\xi = \begin{pmatrix} A & B \\ C & D \end{pmatrix} \begin{pmatrix} q \\ \sigma q \end{pmatrix} = \begin{pmatrix} (A+B\sigma)q \\ (C+D\sigma)q \end{pmatrix} = \begin{pmatrix} q \\ \sigma q' \end{pmatrix} . \tag{10.10}$$

We thus have the action

$$g*\sigma = \sigma' = (C+D\sigma)(A+B\sigma)^{-1} . \tag{10.11}$$

In terms of w:

$$w \to g*w = [A-iB-iC-D + (A+iB-iC+D)w]$$

$$\times [A-iB+iC+D + (A+iB+iC-D)w]^{-1} . \tag{10.12}$$

Before attempting to construct the oscillator representation, let us learn something from the case n = 2. In that case X is homeomorphic to the projective cone in 3+2-dimensional, pseudo-Euclidean space. However, it turns out that it is much more convenient to deal with homogeneous fields on the cone itself, since these transform as ordinary scalar and spinor fields--locally, at least. In fact, it is necessary to pass to a double covering of the cone, and a four-fold covering of X. Projecting down to X one obtains multiplier representations, with multipliers determined by the degrees of homogeneities of the fields. We shall determine the action of SP(2n) on n-forms, the analogue of the 3+2 cone, and obtain from that the required multipliers.

11. The line bundle Z^{α} over X.

An n-form $z \in Z$ is more than the plane $x = \beta(z)$; it also has a (complex) value. The map $z \to \beta(z)$ defined by (9.2) and (9.3) determines z up to a multiplicative constant; therefore Z is a complex line bundle over X, with projection β and fiber $\beta^{-1}(x) = C$. The symplectic group acts in V_{2n} and on z by pullback. We have already calculated the action of SP(2n) in the base manifold; now we must find out how it acts in the fiber.

We claim that the effect of a symplectic transformation on $\xi = q+i0 \in x_o$ is expressible as

$$\begin{pmatrix} q \\ 0 \end{pmatrix} \to \begin{pmatrix} q' \\ p' \end{pmatrix} = \begin{pmatrix} (w+1)Rq \\ i(w-1)Rq \end{pmatrix}, \tag{11.1}$$

where w is unitary and symmetric and $R \in GL(n,C)$. If this is true, then the effect of another symplectic transformation $g \in SP(2n)$ takes the form

$$\begin{pmatrix} q' \\ p' \end{pmatrix} \to \begin{pmatrix} A & B \\ C & D \end{pmatrix} \begin{pmatrix} q' \\ p' \end{pmatrix} = \begin{pmatrix} (g*w+1)\, g\cdot Rq \\ i(g*w-1)\, g\cdot Rq \end{pmatrix} \tag{11.2}$$

where $g{*}w$ is another unitary, symmetric matrix and $g{\cdot}R \in GL(n,C)$. Indeed, we easily verify that $g{*}w$ is precisely as given by (10.12); and that

$$g{\cdot}R = M(g,w)\, R \,, \tag{11.3}$$

where $M(g,w)$ is the complex n-by-n matrix

$$M(g,w) = \frac{1}{2}\,[A-iB+iC+D + (A+iB+iC-D)w] \,. \tag{11.4}$$

Eq. (11.2) defines an action of SP(2n) in $X \times GL(n,C)$. Now consider the complex line bundle over X that is obtained by mapping GL(n,C) onto C by the determinant. We then get an action in the fiber given by

$$[\det g{\cdot}R/\det R]_x = \det M(g,w) \,. \tag{11.5}$$

We use (11.2) to evaluate the action of SP(2n) on the value of an isotropic n-form. Let $\{\tilde{e}^1,\ldots,\tilde{e}^n\}$ be the basis for x_o given by $(\tilde{e}^i)_j = \delta_j^{\ i}$, and let $\{e_1,\ldots,e_n\}$ be the vectors in V^*_{2n} that have components $(e_i)^j = \delta_i^{\ j}$. Let z_o be the n-form

$$z_o = e_1 \wedge \ldots \wedge e_N \,, \quad z_o(\tilde{e}^1,\ldots,\tilde{e}^n) = n! \,. \tag{11.6}$$

Now set

$$\xi^i = \begin{pmatrix} (w+1)\, R\, \tilde{e}^i \\ i(w-1)\, R\, \tilde{e}^i \end{pmatrix}, \quad i = 1,\ldots,n \,, \tag{11.7}$$

and evaluate

$$z_0(\xi^1,\dots,\xi^n) = \varepsilon_{i_1\dots i_n}[(w+1)\,R]_1^{\ i_1}\ \dots\ [(w+1)\,R]_n^{\ i_n} = \det[(w+1)\,R]\ . \tag{11.8}$$

The action of $g^{-1} \in SP(2n)$ in Z is therefore given by

$$g^{-1}\,z(w) = \frac{\det[(g*w+1)\ g\cdot R]}{\det[(w+1)\ R]}\,z(g*w)\ . \tag{11.9}$$

It is convenient to introduce the complex function z^1,

$$z^1(w) = z(w)/\det(w+1)\ , \tag{11.10}$$

in terms of which (11.9) simplifies to

$$g^{-1}z^1(w) = \mu(g^{-1},w)\ z^1(g*w)\ , \tag{11.11}$$

$$\mu(g^{-1},w) = \det M(g,w)\ . \tag{11.12}$$

This is a multiplier representation with multiplier $\mu(g,w)$, provided that the co-cycle equation holds:

$$\mu(gg',w) = \mu(g,w)\ \mu(g',g^{-1}*w)\ . \tag{11.13}$$

That it does, follows from the fact that M(g,w) satisfies the same equation; this in turn is guaranteed by the construction from (11.2).

Since the oscillator representation of sp(2n) exponentiates to a projective representation of SP(2n); that is, to a representation of a covering of this group, we must pass to a covering of Z. This is a subject that is treated exhaustively in geometric quantization. Briefly, one considers a bundle Z^α of α-densities on X. The action of the covering group on a section ϕ is given by

$$g\phi(w) = \mu(g,w)\ \phi(g^{-1}*w)\ , \tag{11.14}$$

where now (11.12) is replaced by

$$\mu(g,w) = [\det M(g^{-1},w)]^{\alpha}\ . \tag{11.15}$$

The co-cycle equation (11.13) holds for any value of α, with g and g' in the universal covering group of SP(2n). It will turn out that $\alpha = -1/2$, and that only the double covering group intervenes. We pass over the precise global definitions and proceed instead to investigate the neighborhood of the identity.

The action of sp(2n) is obtained by substituting

$$g = \begin{pmatrix} A & B \\ C & D \end{pmatrix} \to 1 + \begin{pmatrix} a & b \\ c & d \end{pmatrix} = 1 + m\ ,$$

and retaining the linear part. From (10.12) we get

$$g^{-1}*w - w \to \delta w \tag{11.16}$$

$$= \frac{1}{2}\{(w-1)\ a\ (w+1) - (w+1)\ d\ (w-1) + i(w-1)\ b\ (w-1) + i(w+1)\ c\ (w+1)\}$$

and the associated vector field δ_m defined by

$$\delta_m \phi(w) = \delta(abcd)\ \phi(w) = \frac{d}{dt}\phi(w+t\delta w)\Big|_{t=0}\ . \tag{11.17}$$

Eqs. (11.14) and (11.4) give the infinitesimal co-cycle

$$\mu(g,w) - 1 \simeq \alpha\ \mathrm{tr}[M(g^{-1},w) - 1] \to \tau_m = \tau(abcd)\ , \tag{11.18}$$

$$\tau(abcd) = -\frac{\alpha}{2}\text{tr}\{\alpha\text{-}ib+ic+d + (a+ib+ic-d)w\}\,. \tag{11.19}$$

We shall show, in the next section, that the choice $\alpha = -1/2$ leads to the even oscillator representation of sp(2n).

12. The oscillator representation on $Z^{-1/2}$.

The action (11.14) of SP(2n) gives rise to the following formal representation of sp(2n) by differential operators acting on the sections of Z^{α}:

$$sp(2n) \ni m = \begin{pmatrix} a & b \\ c & d \end{pmatrix} \to \Delta(abcd) = \Delta_m\,,$$

$$\Delta_m \phi = (\delta_m + \tau_m)\,\phi \quad . \tag{12.1}$$

The vector field δ_m was defined by (11.17) and (11.16), and the infinitesimal cocycle τ_m by (11.19). The cocycle equation (11.13) becomes

$$d\tau(m,m') \equiv [\tau_m,\tau_{m'}] + \delta_m\tau_{m'} - \delta_{m'}\tau_m - \tau_{[m,m']} = 0\,, \tag{12.2}$$

which tells us that

$$[\Delta_m,\Delta_{m'}] = \Delta_{[m,m']}\,, \tag{12.3}$$

as it must be. Now

$$\begin{pmatrix} a & b \\ c & d \end{pmatrix} \in sp(2n) \text{ if } {}^t a = -d,\ {}^t b = b,\ {}^t c = c\,; \tag{12.4}$$

$$\begin{pmatrix} a & b \\ c & d \end{pmatrix} \in u(n) \text{ if } a = d = -{}^t a,\ b = -c = {}^t b\,. \tag{12.5}$$

The energy is the u(1) generator

$$h = \frac{1}{2}\begin{pmatrix} 0 & 1 \\ -1 & 0 \end{pmatrix} ; \quad a = d = 0 , \quad b = -c = (1/2)\, 1 . \tag{12.6}$$

Specialization of (11.16) and (11.19) gives:

For u(n):

$$\tau(abcd) = -i\alpha \operatorname{tr} c , \quad \delta w = -(a-ic)\, w - w\, {}^t(a-ic) , \tag{12.7}$$

and $\tau = 0$ on su(n). For the energy

$$i\tau(abcd) = -\alpha n/2 , \quad i\delta w = w . \tag{12.8}$$

For the lowering operators (negative, noncompact roots)

$$\begin{pmatrix} a & ia \\ ia & -a \end{pmatrix} ; \quad b = c = ia , \quad d = -a = -\, {}^t a ,$$

$$\tau(abcd) = 0 , \quad \delta w = -2a . \tag{12.9}$$

For the energy raising operators (positive, noncompact roots)

$$\begin{pmatrix} a & -ia \\ -ia & -a \end{pmatrix} ; \quad b = c = -ia , \quad d = -a = -\, {}^t a ,$$

$$\tau(abcd) = -2\alpha \operatorname{tr}(aw) , \quad \delta w = 2waw . \tag{12.10}$$

Therefore: by (12.9), ϕ = constant is a lowest weight vector; by (12.7) it is su(n) invariant and by (12.8) it has energy $-\alpha n/2$. We thus get the positive energy, even oscillator representation, perhaps as a subquotient of a nondecomposable representation, by choosing $\alpha = -1/2$.

We now show that the module generated from the highest weight vector $\phi = 1$, by repeated application of the operators Δ_m, is in fact irreducible. Of course, it is sufficient to apply the operators associated with the positive, noncompact roots. According to (12.1) and (12.10) these operators are, for $\alpha = -1/2$,

$$\Delta(a) \equiv \Delta(a,-ia,-ia,-a) = \mathrm{tr}(aw+2waw\partial) \; . \tag{12.11}$$

Thus

$$\Delta(a)\; 1 = \mathrm{tr}(aw) \; , \tag{12.12}$$

$$\begin{aligned} \Delta(a')\; \Delta(a)\; 1 &= \mathrm{tr}(a'w + 2wa'w\partial)\; \mathrm{tr}(aw) \\ &= \mathrm{tr}(a'w)\; \mathrm{tr}(aw) - 2\mathrm{tr}(wa'waw) \\ &= w_{ij} w_{k\ell} [a'_{ij} a_{k\ell} + a'_{jk} a_{i\ell} + a'_{ki} a_{j\ell}] \; . \end{aligned} \tag{12.13}$$

These vectors have energy n/4 + 2, and irreducibility can be established here. The lowest weight of the even oscillator representation is $\lambda = (\frac{1}{2},...,\frac{1}{2})$. This is the penultimate reduction point; the reduction occurs in the third level (corresponding to E = n/4 + 2), at $\lambda' = (\frac{5}{2},\frac{5}{2},\frac{1}{2},...,\frac{1}{2})$. The projection of this on su(n) is (2,2,0,...,0). It is enough to show that this representation of su(n) does not occur in (12.13). In fact, since a and a' are symmetric, only the completely symmetric projection of the tensor $w_{ij} w_{k\ell}$ occurs. Therefore, su(n) operates irreducibly on this level, according to the representation with highest su(n) weight (4,0,...,0). This proves that the module generated from $\phi = 1$ is irreducible.

Now it follows that the operators I_{abcd} of Eq. (2.10) vanish on this module, and from Section 4 that I_{ab} vanishes in the space generated from $\phi = 1$ by the enveloping algebra of osp(2n/1). Consequently, we have an irreducible representation of osp(2n/1), equivalent to the oscillator representation. The important difference between this realization and the original one that was obtained by Weyl quantization, is that here the generators are first order differential operators acting as vector fields on a manifold. This should make it possible to formulate a field theory with a local action principle.

To sum up our progress so far, we have identified the homogeneous space that is most appropriate for our purpose, and we have obtained an explicit realization of osp(2n/1). In the case of greatest physical interest, $n = 4$ and $\dim X = 10$. Our most immediate problem is to understand the physical meaning of the 10 dimensions of X.

At the risk of pointing out the obvious, let us note that the space generated from the lowest weight vector (ϕ = constant) is the space of polynomials in the components of the matrix w. More precisely, let D be the bounded symmetric domain of symmetric matrices defined by

$$1 - ww^* \geq 0 \ .$$

The action of SP(2n) on X extends naturally to the space of polynomials on D. This domain will turn out to be important for the physical interpretation. It is analogous to the forward tube of relativistic field theories, as will be seen later.

V. Physical Interpretation

The physical interpretation proceeds from the identification of the conformal group SU(2,2) as a subgroup of SP(8,R). [We actually study the general case of SU(n/2,n/2) in SP(2n,R).] Physical space time is the unique orbit [in the homogeneous space studied above] of SU(2,2) of minimal dimension. All the orbits are determined. The oscillator representation of SP(2n,R) reduces to a sum of massless representations of the conformal group.

The imbedding of physical space time in a space of higher dimension is reminiscent of Kaluza-Klein theories. Here, all spaces are compact from the start, and the harmonic expansion in the extra dimensions is both natural and rigorously justified. In fact, the expansions converge in a bounded symmetric domain that extends the forward tube domain of relativistic field theories.

13. The conformal group $U^\gamma(n)$.

The physical interpretation requires that we identify physical space time, and the physically recognizable representations of the conformal group, in terms of fields on space time. The conformal group is a sub-group of SP(2n) and therefore acts on X. It does not act transitively, however. Space time is a very special orbit of the conformal group in X, of the lowest possible dimension. This turns out to be $(n/2)^2$ in general and 4 in the physically interesting case $n = 4$. In this section we begin the study of the conformal group and determine the orbit of highest dimension. The other orbits, including space time, will be found in the subsequent Sections.

Let γ be a real, symmetric n-by-n matrix, and let h^γ be the symplectic matrix

$$h^{\gamma} = \frac{1}{2}\begin{pmatrix} 0 & \gamma \\ -\gamma & 0 \end{pmatrix} . \tag{13.1}$$

In analogy with the definition of U(n) as the commutant of h, we define

$$U^{\gamma}(n) = \text{commutant of } h^{\gamma} \text{ in } SP(2n) . \tag{13.2}$$

If γ is the unit matrix, then $h^{\gamma} = h$ and $U^{\gamma}(n) = U(n)$. However, from now on we shall suppose that

$$n \text{ is even} , \quad \gamma = \begin{pmatrix} 1 & 0 \\ 0 & -1 \end{pmatrix} , \quad \operatorname{tr} \gamma = 0 ; \tag{13.3}$$

then $U^{\gamma}(n) = U(n/2,n/2)$. The conformal group of 4-dimensional space time is

$$\mathscr{C} = SU(2,2)/Z_2 \otimes Z_2 . \tag{13.4}$$

For simplicity we shall refer to $U^{\gamma}(n)$ as the conformal group.

Just as it was convenient to interpret U(n) as the group of unitary matrices acting in a complex vector space W_n, so it is useful to associate $U^{\gamma}(n)$ with the pseudo-unitary endomorphisms of a complex vector space W_n^{γ}. We make the identification

$$V_{2n} \ni \begin{pmatrix} q \\ p \end{pmatrix} \leftrightarrow q + i\gamma p \in W_n^{\gamma} .$$

Hence W_n^{γ} differs from W_n by the choice of complex structure. The hermitean structure is also determined by γ, so that W_n^{γ} is an indefinite inner product space. If g belongs to $U^{\gamma}(n)$, as defined by (13.2), then the associated n-by-n pseudo-unitary matrix will also be denoted g. Hence

$$U^\gamma(n) \ni g = \begin{pmatrix} A & B \\ C & D \end{pmatrix} \leftrightarrow g = A + i\gamma C \in U^\gamma(n) \; ; \tag{13.5}$$

$$U^\gamma(n) = \{g \in GL(n,c) \; ; \; g\gamma g^\dagger = \gamma\} \; , \tag{13.6}$$

where $g^\dagger$ is the ordinary hermitean conjugate of g.

Now $U^\gamma(n)$ acts in X, but unlike U(n) it does not act transitively. We shall begin by showing that there is a unique orbit that is open and dense in X.

Recall the definition (10.1) of the reference point x_o in X. Let X^γ be the orbit of $U^\gamma(n)$ through x_o, so that

$$X^\gamma = U^\gamma(n)\, x_o = \{gx_o \; ; \; g \in U^\gamma(n)\} \; . \tag{13.7}$$

The stabilizer of x_o in $U^\gamma(n)$ is the real subgroup of (13.6), namely

$$O^\gamma(n) = \{g \in GL(n,R) \; ; \; g\,\gamma\,{}^t g = \gamma\} \; .$$

In view of (13.3), $O^\gamma(n) = O(n/2,n/2)$. Consequently,

$$X^\gamma = U^\gamma(n)/O^\gamma(n) \; ,$$

and this orbit has the same dimension as X. This also shows that we can parameterize the points of X^γ in strict analogy with the parameterization of X by unitary symmetric matrices. If $g \in U^\gamma(n)$, then the vector ξ belongs to the plane gx_o if and only if $\xi^* = g^* g^{-1} \xi$. This establishes a bijection between X and the image of the mapping

$$U^\gamma(n) \ni g \to g^* g^{-1} = \dot{w} \in U^\gamma(n) \; , \tag{13.8}$$

and a parameterization of X^γ.

$$X^{\gamma} = \{\dot{w} \in U^{\gamma}(n)\ ,\ \gamma\dot{w}\gamma = {}^{t}\dot{w}\}\ . \tag{13.9}$$

This is the analogue of (10.8); it should be stressed that these parameterizations are complete and one-valued.

We shall find a simple characterization of X^{γ} and its complement in X. Recall that, if $g \in U(n)$, $w = g^{*}g^{-1}$ and $\xi \in gx_{o}$, then ξ has the representation

$$\xi = \begin{pmatrix} (w+1)\,Rq \\ i(w-1)\,Rq \end{pmatrix}\ .$$

If w+1 is invertible, then

$$\xi = \begin{pmatrix} q' \\ \sigma q' \end{pmatrix}\ ,\quad \sigma = i\,\frac{w-1}{w+1}\ ,\quad w = \frac{i+\sigma}{i-\sigma}\ , \tag{13.10}$$

where σ is real and symmetric. On the other hand, if $\beta \in U^{\gamma}(n)$, $\dot{w} = \beta^{*}\beta^{-1}$ and $\xi \in \beta x_{o}$, then

$$\xi = \begin{pmatrix} (\dot{w}+1)\,\dot{R}q \\ i\gamma(\dot{w}-1)\dot{R}q \end{pmatrix}\ ,$$

and if $\dot{w}+1$ is invertible

$$\xi = \begin{pmatrix} \dot{q}' \\ \gamma\dot{\sigma}q' \end{pmatrix}\ ,\quad \dot{\sigma} = i\,\frac{\dot{w}-1}{\dot{w}+1}\ ,\quad \dot{w} = \frac{i+\dot{\sigma}}{i-\dot{\sigma}}\ . \tag{13.11}$$

By (13.10) and (13.11) we have a bijection between certain dense subsets of X and X^{γ}, given by

$$\sigma = \gamma\dot{\sigma}\ . \tag{13.12}$$

This correspondence is initially valid under the condition that w+1, $\dot{w}+1$ and $i-\dot{\sigma}$ be invertible. (Since σ is symmetric, $i-\sigma$ is always invertible.)

To extend it, we write (13.12) in terms of w and $\dot{w}$:

$$[\gamma + 1 + w(\gamma-1)]\,\dot{w} = \gamma - 1 + w(\gamma+1)\ , \tag{13.13}$$

$$w\,[\gamma + 1 + (1-\gamma)\dot{w}] = 1 - \gamma + (\gamma+1)\dot{w}\ . \tag{13.14}$$

These are just two ways of writing the same equality.

Let us put

$$\dot{w} = \begin{pmatrix} v & u \\ -{}^tu & \bar{v} \end{pmatrix}\ , \text{ v and } \bar{v} \text{ symmetric}\ ,$$

then (13.14) reads

$$w\begin{pmatrix} 1 & 0 \\ -{}^tu & \bar{v} \end{pmatrix} = \begin{pmatrix} v & u \\ 0 & 1 \end{pmatrix}. \tag{13.15}$$

Now $\dot{w}$ is pseudo-unitary, $\dot{w}\gamma\dot{w}^\dagger = \gamma$ or $\dot{w}\dot{w}^* = 1$. Explicitly,

$$\begin{pmatrix} vv^* - uu^\dagger\ , & vu^* + u\bar{v}^* \\ -{}^tuv^* - \bar{v}u^\dagger\ , & \bar{v}\bar{v}^* - {}^tuu^* \end{pmatrix} = 1\ .$$

This shows that $\bar{v}$ is always invertible, so that Eq. (13.15) can always be solved for w. Therefore, Eq. (13.14 extends (13.12) to an injection of X^γ into X. On the other hand, put

$$w = \begin{pmatrix} v & u \\ {}^tu & \bar{v} \end{pmatrix}\ , \text{ v and } \bar{v} \text{ symmetric}\ , \tag{13.16}$$

then (13.13) reads

$$\begin{pmatrix} 1 & -u \\ 0 & -\bar{v} \end{pmatrix} \dot{w} = \begin{pmatrix} v & 0 \\ {}^tu & -1 \end{pmatrix} . \tag{13.17}$$

Unitarity of w is expressed by

$$\begin{pmatrix} vv^*+uu^\dagger \, , & vu^*+u\bar{v}^* \\ {}^tuv^*+\bar{v}u^\dagger \, , & \bar{v}\bar{v}^*+{}^tuu^* \end{pmatrix} = 1 . \tag{13.18}$$

This shows that the obstruction to the extension of (13.12) to all of X is the non-invertibility of $\bar{v}$. Note that v and $\bar{v}$ have the same rank, and that we have the identities

$$(w+1)\gamma(1+\dot{\sigma}^2)(w+1) = 2(w\gamma+\gamma w) = 4\begin{pmatrix} v & 0 \\ 0 & -\bar{v} \end{pmatrix} . \tag{13.19}$$

We summarize as follows.

Theorem. The complement $X - X^\gamma$ of the unique open orbit X^γ of $U^\gamma(n)$ in X consists of the points where $\gamma w+w\gamma$ is not invertible.

14. Identification of space time.

The action of $U^\gamma(n)$ is not transitive in $X - X^\gamma$. This exceptional submanifold consists of several orbits of different dimensions. Space time is the orbit of lowest dimension; it will be determined in this section. The other orbits are also of some interest and we shall study them in the next section.

To begin with, we use the parameterization of X in terms of real, symmetric matrices. The action of SP(2n) on the real, symmetric matrix σ was given by Eq. (10.11), namely

$$g{*}\sigma = (C+D\sigma)(A+B\sigma)^{-1}\ ,\quad g = \begin{pmatrix} A & B \\ C & D \end{pmatrix}\ . \tag{14.1}$$

The corresponding infinitesimal action is expressed by

$$\delta\sigma = -\sigma a - \sigma b\sigma + d\sigma + c\ ,\quad g = 1 + \begin{pmatrix} a & b \\ c & d \end{pmatrix}\ . \tag{14.2}$$

We are interested in the action of $U^{\gamma}(n)$ and note that

$$\begin{pmatrix} a & b \\ c & d \end{pmatrix} \in u^{\gamma}(n) \Longleftrightarrow a = \gamma d\gamma = -\,{}^{t}d\ ,\quad b = {}^{t}b = -\gamma c\gamma\ . \tag{14.3}$$

Now choose a fixed point σ, and vary (abcd) in $u^{\gamma}(n)$. If the orbit of $U^{\gamma}(n)$ through σ is of dimension less than that of X, then there exists a space $\mathcal{N}_{\sigma}$ of vectors that are normal to the orbit at σ. More precisely, $\mathcal{N}_{\sigma}$ is a space of linear functionals on the tangent space of X at σ; it consists of all the real, symmetric matrices N, such that

$$\mathrm{tr}(N\delta\sigma) = 0\ ,\quad \begin{pmatrix} a & b \\ c & d \end{pmatrix} \in u^{\gamma}(n)\ . \tag{14.4}$$

Explicitly, this condition reads

$$\mathrm{tr}[a(\gamma\sigma N\gamma - N\sigma) + c(N + \gamma\sigma N\sigma\gamma)] = 0\ .$$

The matrices a and c are arbitrary except for the conditions of symmetry implied by (14.3); that is, γa is antisymmetric and c is symmetric. Now it happens that the matrices that multiply a and c have the same symmetries; so they must vanish, and we end up with the following two conditions on N:

$$[\dot{\sigma}, N\gamma] = 0\ ,\quad (1+\dot{\sigma}^2)\,N = 0\ , \tag{14.5}$$

with $\dot{\sigma} = \gamma\sigma$.

The co-dimension of the $U^{\gamma}(n)$ orbit through σ is equal to the dimension of $\mathcal{N}_{\sigma}$, and (14.5) shows that it depends strongly, if not exclusively, on the multiplicity of -1 in the spectrum of $\dot{\sigma}^2$. If -1 is not in the spectrum of $\dot{\sigma}^2$, then dim $\mathcal{N}_{\sigma} = 0$. The collection of all such points is the orbit X^{γ} of maximal dimension. The complement of X^{γ} in X consists of orbits of lower dimension.

The $U^{\gamma}(n)$ orbit of minimal dimension consists of points where $\mathcal{N}_{\sigma}$ is of maximal dimension; that is, the points at which $\dot{\sigma}^2 = -1$. This orbit will be referred to as space time and denoted X_o:

$$\text{space time} = X_o = \{x \in X\,;\ \dot{\sigma}^2 = -1\}\,. \tag{14.6}$$

To make this quite precise, we should use the global parameterization in terms of unitary, symmetric matrices. The correct definition is

$$\text{space time} = X_o = \{x \in X\,;\ w\gamma+\gamma w = 0\}\,. \tag{14.7}$$

Here w is the matrix that corresponds to the plane x when X is parameterized by unitary, symmetric matrices of order n, while $\dot{\sigma} = \gamma\sigma$ and σ is the matrix associated with x in the parameterization of X by real, symmetric matrices.

In view of the particular form (13.3) chosen for the matrix γ, we see that w belongs to X_o if and only if

$$w = \begin{pmatrix} 0 & u \\ {}^t u & 0 \end{pmatrix}, \quad u \in U(n/2)\,. \tag{14.8}$$

Therefore, we have the final result

$$\text{space time} = X_o = U(n/2)\,. \tag{14.9}$$

In the physically interesting case n = 4 we obtain U(2), which is precisely the conformal compactification of Minkowski space. In this case the dimension of X_0 is 4; in the general case it is $(n/2)^2$.

We record here the action of the conformal group $U^\gamma(n)$ on space time. It can be calculated from (10.12), or more easily from (11.16). The result, in the notation of (13.5), is

$$U^\gamma(n) \ni A + i\gamma C = g = \begin{pmatrix} p & q \\ r & s \end{pmatrix}^* , \tag{14.10}$$

$$g*u = (pu+q)/(ru+s) . \tag{14.11}$$

The complex conjugation in the last member of (14.10) is introduced to keep it out of (14.11). For the co-cycle we obtain

$$\mu_0(g^{-1},u) = [\det(ru+s)]^{-1} . \tag{14.12}$$

This is obtained from the multiplier $\mu(g^{-1},w)$ by setting $v = \bar{v} = 0$ and choosing g as in (14.10). The restriction of (11.4) to $g \in U^\gamma(n)$ and $w \in$ space time is thus

$$g^{-1}\phi(u) = [\det(ru+s)]^{-1} \phi(g*u) . \tag{14.13}$$

15. The other orbits of $U^\gamma(n)$.

We continue our study of the space $\mathcal{N}_\sigma$ of normals to the orbit at σ, in the case when w+1 is invertible. To simplify the algebra set

$$N = (w+1)\, M\, (w+1) , \tag{15.1}$$

then (14.5) become

$$(w+1)\, M\, (w-1) = \gamma(w-1)\, M\, (w+1)\, \gamma ,$$

$$(w+1)\, M\, (w+1) = \gamma(w-1)\, M\, (w-1)\, \gamma .$$

If w+1 is invertible, then we easily deduce that

$$M(w\gamma+\gamma w) = 0 , \tag{15.2}$$

$$(1-w)[M,\gamma] = 0 . \tag{15.3}$$

We have seen that X^γ consists of points where $w\gamma+\gamma w$ is invertible. We now place ourselves in $X - X^\gamma$ by postulating that it is not. Thus, in the notation of Section 13:

$$w = \begin{pmatrix} v & u \\ {}^t u & \overline{v} \end{pmatrix} , \quad \gamma w + w\gamma = 2\begin{pmatrix} v & 0 \\ 0 & -\overline{v} \end{pmatrix} , \tag{15.4}$$

where v and $\overline{v}$ are symmetric matrices of order n/2 and rank $q < n/2$. Recall that v and $\overline{v}$ have equal rank, since w is unitary.

The matrices that belong to $U(n) \cap U^\gamma(n)$ have the form

$$\begin{pmatrix} \alpha & 0 \\ 0 & \beta \end{pmatrix} , \quad \alpha,\beta \in U(n/2) . \tag{15.5}$$

When w is expanded as in (15.4), then the action of this matrix on w amounts to

$$u \to \alpha\, u\, {}^t\beta , \quad v \to \alpha\, v\, {}^t\alpha , \quad \overline{v} \to \beta\, \overline{v}\, {}^t\beta . \tag{15.6}$$

By means of a transformation of this type we can bring $u,v,\overline{v}$ to the form

$$u = \begin{pmatrix} \mu & 0 \\ 0 & \mu' \end{pmatrix}, \quad v = \begin{pmatrix} 0 & 0 \\ 0 & v' \end{pmatrix}, \quad \bar{v} = \begin{pmatrix} 0 & 0 \\ 0 & \bar{v}' \end{pmatrix}. \tag{15.7}$$

The lower right blocks are square matrices of dimension $q = \mathrm{rank}(v) = \mathrm{rank}(\bar{v})$ and μ is unitary.

We are now ready to analyze (15.2) and (15.3). It is easy to see that these two equations imply $[M,\gamma] = 0$, even in the case when $1-w$ is singular. Therefore, M must have the form

$$M = \begin{pmatrix} M_1 & 0 \\ 0 & M_2 \end{pmatrix}, \quad M_i = \begin{pmatrix} m_i & 0 \\ 0 & 0 \end{pmatrix}, \quad i = 1,2 .$$

Here the dimensions of M, M_i, and m_i are n, $n/2$, $(n/2)-q$. Now Eq. (15.1) becomes

$$N = \begin{bmatrix} m_1 + \mu\, m_2\, {}^t\mu & , 0 & , m_1\, \mu + \mu\, m_2 & , 0 \\ 0 & , 0 & , \quad 0 & , 0 \\ {}^t\mu\, m_1 + m_2\, {}^t\mu & , 0 & , {}^t\mu\, m_1\, \mu + m_2 & , 0 \\ 0 & , 0 & , \quad 0 & , 0 \end{bmatrix}$$

Recall that N is a normal to the orbit in the sense that $\mathrm{tr}(N\delta\sigma)$ vanishes. We could choose the transformation (15.5) so as to make $\mu = 1$, but in that case $w+1$ is not invertible and the parameterization by the real, symmetric matrices σ breaks down. Instead we may go to the point $\mu = i$, where $w+1$ is invertible. Then it is easy to impose the condition that N be real, namely $m_1 + m_2^\dagger = 0$, which leaves us with $(n/2 - q)(n/2 - q + 1)$ free, real parameters in N. This is the co-dimension of the orbit.

Theorem. (i) The dimension of the $U^\gamma(n)$ orbit through any point w at which rank $(\gamma w+w\gamma) = 2q$ is

$$\frac{n}{2}(n+1) - \left(\frac{n}{2} - q\right)\left(\frac{n}{2} - q + 1\right).$$

(ii) For each value of q there is precisely one orbit; it will be called the orbit of rank q and denoted X_q.

The second statement is justified as follows. The main point is that we have determined all the normals to the orbit. Therefore $U^\gamma(n)$ acts transitively on the submatrices $\mu,\mu',v',\bar{v}'$ in (15.7), within the domain given by $ww^* = 1$. In the case n = 4 we have precisely three orbits:

(i) X_0 = space time, dimension 4,

$$\gamma w + w\gamma = 0\,, \quad 1 - uu^\dagger = 0\,; \tag{15.8}$$

(ii) X_1, dimension 8,

$$\mathrm{rank}(\gamma w+w\gamma) = 2\,, \quad \det(1-uu^\dagger) = 0\,; \tag{15.9}$$

(iii) $X_2 = X^\gamma$, dimension 10,

$$\gamma w + w\gamma \text{ invertible}\,, \quad 1-uu^\dagger > 0\,. \tag{15.10}$$

Let us return to the general case.

The space D_0 of (n/2)-dimensional, square, complex matrices defined by

$$D_0\!: \; 1-uu^\dagger > 0 \tag{15.11}$$

is one of the classical, bounded symmetric domains of E. Cartan. It is the symmetric space G/K, where G = U(n/2,n/2) and K is the maximal compact subgroup. The characteristic manifold of D_o is the portion of its boundary that is defined by $uu^\dagger = 1$; that is, space time. Hence D_o is a complexification of space time; it is related to the forward tube that plays an important part in the formulation of quantum field theory. When n = 4, then it happens that the dimensions of D_o and of X_1 are equal, but these two spaces are not equivalent as homogeneous spaces for $U^\gamma(4)$. The stabilizer of X_1 is not semisimple. Another important classical domain is the complexification of X given by

$$D:\ \ 1-ww^* > 0\ . \tag{15.12}$$

Its characteristic manifold is X. Both of these domains are important for the physical interpretation of the fields defined on X.

16. Interpretation of the extra dimensions.

Let n = 4 and let us imagine a system of coordinates (u,v) = $(u^1,...,u^4;v^1,...,v^6)$ for X, so that space time is the submanifold $v^1 = ... = v^6 = 0$. Then we can conceive of a Taylor series development of the form

$$\phi(w) = \phi(u,v) = \sum_k \phi_{\alpha_1...\alpha_k}(u)\, v^{\alpha_1} \ldots v^{\alpha_k}\ . \tag{16.1}$$

This series might define a sequence of tensors of order k = 0,1,2,..., depending only on u, and such space time tensor fields could be associated with the fields of electromagnetism, gravity, etc.

We found the even oscillator representation of sp(2n) realized on a space of polynomials in the matrix elements of w; this realization does not depend on w being unitary, and it extends naturally to a space of holomorphic functions on the disc (15.12). Let $\mathcal{A}_o(D)$ be this space; it has a basis that consists of homogeneous polynomials of the form

$$\Sigma\, w_{ij} w_{k\ell} \cdots , \tag{16.2}$$

where the summation is over all permutations of the indices. Now extend the parameterization (13.16) of X to D. Then $\phi \in \mathcal{A}_o(D)$ is analytic in the (n/2)(n+1) complex variables $u,v,\bar{v}$ and $\phi(u,...)$ is analytic in $v,\bar{v}$. There is a unique expansion

$$\phi(u,v,\bar{v}) = \sum_{k,\ell} \phi^{(k\alpha)(\ell\beta)}(u)\, v_{(k\alpha)}\bar{v}_{(\ell\beta)} ,$$

$$(k\alpha) = \alpha_1,...,\alpha_k \, ; \quad v_{(k\alpha)} = v_{\alpha_1} \cdots v_{\alpha_k} , \tag{16.3}$$

where each Greek index stands for a pair; e.g. $\alpha_1 = i_1 j_1$, with i_1 and j_1 running from 1 to n/2. Hence ϕ is determined in D (and on X) by the coefficients efficients $\phi^{(k\alpha)(\ell\beta)}$. These coefficients are functions of u, analytic in the domain of complex matrices u such that there is a w of the form (13.16) in D. This is just the domain D_o defined by (15.11).

Holomorphic functions on D_o are fixed by their values on the characteristic manifold; that is, space time. The restriction of $\{\phi^{(k\alpha)(\ell\beta)}\}$ to space time thus determines ϕ as a holomorphic function on D, and the values of ϕ on X are boundary values of this analytic function.

To summarize: The even oscillator representation was realized on a space of functions on X. These functions are the boundary values of holomorphic functions that are determined by a sequence of fields on space time. Therefore, the oscillator representation can be realized in terms of these fields on space time.

We now study the restriction of the oscillator representation to the conformal group $SU^{\gamma}(n) = SU(n/2,n/2)$. Recall that the subgroups $U(1)$ and $SU(n/2,n/2)$ of $U^{\gamma}(n) = U(n/2,n/2)$ sit as a dual pair (in the sense of R. Howe) inside the oscillator representation. This means that the restriction to $U^{\gamma}(n)$ takes the form

$$\text{(even osc.)}\Big|_{U(1)\otimes SU(n/2,n/2)} = \sum_{s=0,\pm 1,\dots} \chi_s \otimes D_s^{\gamma} . \tag{16.4}$$

Here χ_s is a unitary character of $U(1)$ and D_s^{γ} is an irreducible representation of $SU^{\gamma}(n) = SU(n/2,n/2)$. Furthermore, D_s^{γ} is a "highest" weight representation with minimum energy $|s| + n/4$; that is, a positive energy, massless representation of the conformal group. We shall now demonstrate that the expansion (16.3) explicitly displays the complete reduction (16.4).

According to (12.1), (12.7) and (13.1), the generator h^{γ} of the center of $u^{\gamma}(n)$ is represented by the operator

$$H^{\gamma} = \mathrm{tr}(v\partial_v - \bar{v}\partial_{\bar{v}}) . \tag{16.5}$$

The eigenspace of H^{γ} with eigenvalue s, and associated with the character χ_s of $U(1)$, consists of those summands in (16.3) that have the fixed value s of $k-\ell$. We refer to s as the helicity of D_s^{γ}, and to $s = k-\ell$ as the helicity of the field $\phi^{(k\alpha)(\ell\beta)}$.

We now work out the action of $u^{\gamma}(n)$ on the variables $u,v,\bar{v}$; this is tantamount to showing the action on the coefficient fields. With the notation of (12.4), (12.5) and (14.2) we apply (11.16) and (11.19) to the various special cases. The matrices $f,\dots,h$ have dimension $n/2$, f and $\bar{f}$ are skew, g and $\bar{g}$ are symmetric.

For $u^\gamma(n) \cap u(n)$,

$$a = d = \begin{pmatrix} f & 0 \\ 0 & \bar{f} \end{pmatrix}, \quad b = -c = \begin{pmatrix} g & 0 \\ 0 & \bar{g} \end{pmatrix}, \tag{16.6}$$

$$\delta u = u(\bar{f}-i\bar{g}) - (f+ig)\, u\,, \tag{16.7}$$

$$\delta v = v(f-ig) - (f+ig)\, v\,, \tag{16.8}$$

$$\delta \bar{v} = \bar{v}(\bar{f}-i\bar{g}) - (\bar{f}+i\bar{g})\, \bar{v}\,, \tag{16.9}$$

and $2i\tau = \mathrm{tr}(g+\bar{g})$. For the energy raising operators in $u^\gamma(n)$ use (12.10) with

$$a = \begin{pmatrix} 0 & h \\ {}^t h & 0 \end{pmatrix}, \tag{16.10}$$

$$\delta u/2 = u\, {}^t h\, u + v\, h\, \bar{v}\,, \tag{16.11}$$

$$\delta v/2 = v\, h\, {}^t u + u\, {}^t h\, v\,, \tag{16.12}$$

$$\delta \bar{v}/2 = \bar{v}\, {}^t h\, u + {}^t u\, h\, \bar{v}\,, \tag{16.13}$$

and $\tau = 2\,\mathrm{tr}({}^t h\, u)$. For energy lowering operators in $u^\gamma(n)$ use (12.9) with the same matrix a, to get

$$\delta u = -2h\,, \quad \delta v = \delta \bar{v} = 0\,, \quad \tau = 0\,. \tag{16.14}$$

We see that the action of $u^\gamma(n)$ preserves the helicity $s = k-\ell$. The individual degrees of homogeneity k and ℓ in the variables v and $\bar{v}$ are almost preserved, the only offending term being the last term in (16.11).

It is important to understand the meaning of this term. It is evident that the subspace of all polynomials of the form $v_\alpha \bar{v}_\beta \phi^{\alpha\beta}(u,v,\bar{v})$, where $\phi^{\alpha\beta}$ are polynomials, is $u^\gamma(n)$ invariant. However, this subspace is empty; there are no such functions in the representation space $\mathcal{A}_0(D)$. This is related to the extremely reduced K-structure of the oscillator representation, and it is revealed by the fact that the highly symmetrized polynomials (16.2) span the representation space. As far as $u^\gamma(n)$ is concerned, we get simpler realizations of the representations D_s^γ by neglecting the last term in (16.11); but this does not constitute restriction to a subspace, nor passage to a quotient, but merely a bijection of a space of polynomials onto another space of polynomials. We can drop this term if (i) our interest is restricted from sp(2n) to $u^\gamma(n)$ and (ii) we do nothing to deform the representations by introducing interactions.

Transferring the action of $u^\gamma(n)$ to the fields, we see that $\phi^{(k\alpha)(0)}$ and $\phi^{(0)(\ell\beta)}$ transform independently of all the others and thus seem to constitute an invariant subspace for $u^\gamma(n)$. However, $\phi^{(k\alpha)(0)}$ carries D_k^γ and $\phi^{(0)(\ell\beta)}$ carries $D_{-\ell}^\gamma$, so these fields exhaust the representation space. Therefore, the "interior" fields, those with $k\ell > 0$, are not independent. They can all be expressed in terms of the fields "on the edge," those with $k = 0$ or $\ell = 0$. Again, under the provisos (i) and (ii) of the preceding paragraph, we can manage without these "auxiliary fields." The introduction of this last term suggests that these fields play a role similar to the auxiliary fields in supersymmetric field theories. They will be needed to make the algebra close off-shell, which is a prerequisite for introducing interactions.

17. Irreducible $u^\gamma(n)$ modules.

The simplest $u^\gamma(n)$ module is D_0^γ; it can be defined, in view of (16.3) and the discussion of Section 16, by the formula

$$\hat{\phi}(u) = \phi^{(0)(0)}(u) = \phi(w)\Big|_{v=\bar{v}=0} \,. \tag{17.1}$$

Here and below ϕ always runs over the oscillator module; that is, the space $\mathcal{A}_0(D)$ of holomorphic functions on D spanned by (16.2). The action of $u^\gamma(n)$ on $\hat{\phi}$ is given by (12.1), with δw and τ as given by (16.6)-(16.14). The variables v and $\bar{v}$ can be replaced by zero at any time during the application of these formulas. We shall study the scalar field $\hat{\phi}$ in great detail in the next section.

The modules D_s^γ, $s = \pm1,\pm2,\ldots$ can be defined similarly. For $s > 0$ it is the space of coefficients

$$A_{\alpha_1\ldots\alpha_s}(u) = A^{(s\alpha)(0)}(u) = \frac{1}{s!}\frac{\partial}{\partial v_{\alpha_1}}\cdots\frac{\partial}{\partial v_{\alpha_s}}\phi(w)\Big|_{v=\bar{v}=0} \tag{17.2}$$

and for $s < 0$ interchange the roles of v and $\bar{v}$. The trouble with this definition is that it lacks elegance. The coefficients have messy transformation properties, they are not $u^\gamma(n)$ tensor fields, and this makes them very cumbersome, as well as physically unrecognizable.

In order to make contact with conventional massless field theory we must describe the irreducible $u^\gamma(n)$ modules in terms of tensor fields. A simple and direct way to do that is to exploit the fact that the adjoint action of $u^\gamma(n)$ on sp(2n,C) is finite dimensional. It is a direct sum of four irreducible representations, on subspaces of sp(2n,C) defined by

(i) span of h^γ, dimension 1 ,

(ii) $su^\gamma(n)$, dimension n^2-1 ,

(iii) $P_+^\gamma = \{z \in sp(2n,C) \ , \ i[h^\gamma,z] = z\}$,

(iv) $P_-^\gamma = \{z \in sp(2n,C) \ , \ i[h^\gamma,z] = -z\}$. (17.3)

The last two each have dimension (n/2)(n+1). Now let $\Delta(z)$ be the representative of z in the oscillator representation, as defined by (12.1), and define the tensor fields

$$F_z^{\pm}(u) = \Delta(z)\, \phi(w)\Big|_{v=\bar{v}=0} \ , \quad z \in P_\pm^\gamma \ . \tag{17.4}$$

We shall calculate these explicitly and verify that they are tensor modules equivalent to $D_{\pm 1}^\gamma$.

Note that the restriction to $v = \bar{v} = 0$ commutes with the action of $u^\gamma(n)$. The action of $m \in u^\gamma(n)$ on (17.4) is

$$\begin{aligned} \Delta'(m)\, F_z^{\pm}(u) &\equiv \Delta(z)\, \Delta(m)\, \phi(w)\Big|_{v=\bar{v}=0} \\ &= \Delta m\, F_z^{\pm}(u) + [\Delta(z), \Delta(m)]\, \phi(w)\Big|_{v=\bar{v}=0} \\ &= \Delta(m)\, F_z^{\pm}(u) + F^{\pm}_{[z,m]}(u) \ . \end{aligned} \tag{17.5}$$

This displays the fact that $F^\pm$ transforms like tensors; that is to say, as submodules of direct products of a finite representation ($P_\pm^\gamma$) and an infinite one (D_o^γ).

We have

$$\begin{pmatrix} a & b \\ c & d \end{pmatrix} = z \in P_+^{\gamma} \Longleftrightarrow a = ib\gamma = i\gamma c = -\gamma d\gamma \,, \tag{17.6}$$

with b symmetric. In this case (11.16) and (11.19) give

$$\delta w = (i/2)\{(1-\gamma)\, b\, (1-\gamma) - w\, (1+\gamma)\, b\, (1-\gamma)$$

$$- w\, (1-\gamma)\, b\, (1+\gamma) + w\, (1+\gamma)\, b\, (1+\gamma)\} \tag{17.7}$$

$$\tau = (i/4)\, \mathrm{tr}\{b\, (1+\gamma)\, w\, (1+\gamma)\}\,. \tag{17.8}$$

When $v = \overline{v} = 0$, then $\tau = 0$ and

$$\delta u = \delta v = 0\,, \quad \delta\overline{v} = 2i({}^t u\, z_1\, u - {}^t u\, z_2 - z_3\, u + z_4) \tag{17.9}$$

$$b = \begin{pmatrix} z_1 & z_2 \\ z_3 & z_4 \end{pmatrix} = {}^t b\,. \tag{17.10}$$

Hence, finally,

$$F_z^{\,+}(u) = -2\, \mathrm{tr}\{({}^t u\, z_1\, u - {}^t u\, z_2 - z_3\, u + z_4)\, \partial_{\overline{v}}\}\, \phi(w)\Big|_{v=\overline{v}=0}\,. \tag{17.11}$$

This identifies F^+, as a $u^{\gamma}(n)$ module, with the terms that are linear in $\overline{v}$ in (15.3); that is with D^{γ}_{-1}. Similarly F^- carries D_1^{γ}. The generalization to D_s^{γ} is obvious.

The advantage of (17.4) and its generalizations over (17.2) is the simplicity of (17.5) and all that this brings with it. One pays by carrying along a surfeit of components. To make this explicit, let us display the four groups of components F^+:

$$F_z^+ = \sum_{i=1}^{4} \operatorname{tr}(z_i F_i^+) . \tag{17.12}$$

Eq. (17.11) shows that they are not independent; we have the following "subsidiary conditions"

$$F_1^+ = -u\, F_2^+ = -F_3^+\, {}^t u = u\, F_4^+\, {}^t u . \tag{17.13}$$

These restrictions project $P_+^\gamma \otimes D_o^\gamma$ on D_{-1}^γ.

VI. Scalar Field on Space Time

This Part is a digression in which we study the theory of a scalar, conformal field theory. This theory is best known in its Minkowski space formulation. It is more elegant in Dirac's six-cone notation. For us, the most instructive formulation is the one that is the least familiar; it views space time as the group manifold of U(2), and as the distinguished boundary of one of the classical, bounded symmetric domains. This is very relevant for quantum field theory, as can be seen from the fact that this symmetric domain is precisely the forward tube in the complex extension of Minkoswki space. We study the connections that exist between wave operators, Bergman and Cauchy kernels, and Lagrangians. The ultimate purpose is to prepare for an assault on the more formidable problem of formulating an osp(8) invariant field theory.

18. Scalar field on Dirac's projective cone.

In this and the following sections we shall make a very detailed study of the conformal scalar field. We begin with the familiar formulation in Minkowski's notation, transform to Dirac's manifestly covariant six-cone formalism, and finally adopt the interpretation of space time as the group manifold of U(2).

The conformally invariant theory of a scalar, self-interacting field is based on the Lagrangian

$$\mathcal{L} \propto \int d^4x\, [\hat{\phi}\, \partial_x^{\,2}\, \hat{\phi} + \lambda\hat{\phi}^4]\,. \tag{18.1}$$

It is manifestly invariant only under Poincaré transformations. It is also invariant under the transformations of the conformal group $\mathscr{C}$, but we need not carry out a demonstration of this well known fact. Instead, we shall consider only the one parameter subgroup of dilatations.

The group of dilatations acts in Minkowski space by

$$x_\mu \to \theta x_\mu\,, \quad \mu = 0,1,2,3\;; \quad 0 < \theta < \infty\,. \tag{18.2}$$

This action is lifted into the complex, one-dimensional vector space in which $\hat{\phi}$ takes its value, and one has a representation given by

$$T_\theta \hat{\phi}(x) = \theta^{-1}\hat{\phi}(x/\theta)\,. \tag{18.3}$$

Invariance of the interaction term in $\mathcal{L}$ is proved by

$$T_\theta \int d^4x\, \hat{\phi}^4(x) = \int d^4x\, \theta^{-4}\, \hat{\phi}^4(x/\theta) = \int d^4y\, \hat{\phi}^4(y)\,. \tag{18.4}$$

Here we changed variables by setting $y_\mu = x_\mu/\theta$ and used the obvious fact that

$$J = \left|\frac{\partial x}{\partial y}\right| = \theta^4 . \tag{18.5}$$

Invariance of the kinetic term in $\mathcal{L}$ is shown in the same way.

The conformal group does not act globally on Minkowski space (it sends some points to infinity), so we follow Dirac and Penrose in adding some points at infinity, turning it into a compact homogeneous space. In R^6, endowed with a pseudo-Euclidean metric and coordinates y_α, $\alpha = 0,\dots,5$, consider the cone

$$y^2 \equiv \delta^{\alpha\beta} y_\alpha y_\beta \equiv y_0^2 - \vec{y}^2 - y_4^2 + y_5^2 = 0 , \tag{18.6}$$

with $\vec{y}^2 = y_1^2 + y_2^2 + y_3^2$. This equation defines, among other things, the pseudo-Euclidean metric $(\delta^{\alpha\beta}) = (\delta_{\alpha\beta})$. Dirac's projective cone is the four- dimensional space that is obtained when one identifies λy with y for all real, non-zero λ. The group SO(4,2) acts naturally in R^6:

$$y \to \Lambda y . \tag{18.7}$$

Here Λ is pseudo-orthogonal with respect to the metric δ, therefore SO(4,2) acts in the cone and also in the projective cone. In fact, only $SO(4,2)/Z_2$ acts effectively in the projective cone. Minkowski's coordinates are defined by

$$x_\mu = y_\mu/(y_4+y_5) , \quad \mu = 0,1,2,3 . \tag{18.8}$$

The action of SO(4,2) in R^6 induces on these coordinates the classical action of $\mathcal{C}$ in Minkowski space.

Consider a scalar field Φ on the cone. To the action of SO(4,2) on the cone we can associate the operator

$$T_\Lambda \tilde{\phi}(y) = \tilde{\phi}(\Lambda^{-1} y) . \tag{18.9}$$

If we seek an invariant wave operator, then there is the obvious candidate

$$\partial_y^{\;2} = \delta_{\alpha\beta}(\partial/\partial y_\alpha)(\partial/\partial y_\beta) .$$

However, this is not intrinsic on the cone, unless $\tilde{\phi}$ is homogeneous of degree -1 in the y. We therefore suppose that this is the case. A field on Minkowski space is then given by

$$\hat{\phi}(x) = (y_4+y_5)\, \tilde{\phi}(y) . \tag{18.10}$$

Now it turns out that

$$\partial_x^{\;2}\, \hat{\phi}(x) = (y_4+y_5)^3\, \partial_y^{\;2}\, \tilde{\phi}(y) . \tag{18.11}$$

One therefore suspects that the Lagrangian of Eq. (18.1) takes on the manifestly invariant form

$$\mathcal{L} \propto \int (dy)[\tilde{\phi}\, \partial_y^{\;2}\, \phi + \lambda \tilde{\phi}^4] . \tag{18.12}$$

However, there is a catch.

The difficulty is that the projective cone has no invariant measure, though it does have a quasi-invariant measure. Beginning with the invariant Lebesgue measure on R^6, and the invariant δ-function associated with the cone, consider

$$dy = 2\, d^6 y\, \delta(y^2) . \tag{18.13}$$

We introduce a set of intrinsic variables t,y by the polar decompositions

$$y_5 + iy_0 = Y\,e^{it}\ , \quad (\vec{y},y_4) = r\hat{y}\ . \tag{18.14}$$

Here Y, r, t are real and $\hat{y}$ is a four-dimensional unit vector. If $d\hat{y}$ denotes the usual measure on the 3-sphere:

$$dy = 2Y\,dY\,r^3 dr\,dt d\hat{y}\ \delta(Y^2-r^2) = r^3 dr\,dt d\hat{y}\ . \tag{18.15}$$

Now let $\tilde{L}$ be a scalar field on the cone, with a Mellin representation of the form

$$\tilde{L}(y) = \int_R d\rho\ r^{i\rho-4}\ L_\rho(t,\hat{y})\ , \tag{18.16}$$

then the following integral is invariant:

$$\int dy\ \tilde{L}(y) = \int dt\ d\hat{y} \int d\rho \int \frac{dr}{r}\ r^{i\rho}\ L_\rho(t,\hat{y}) \propto \int dt\ d\hat{y}\ L_0(t,\hat{y})\ .$$

This tells us that if $\tilde{L}_0(y)$ is a scalar field on the cone, and if its degree of homogeneity is -4, then the following integral is invariant

$$\int dt\ d\hat{y}\ L_0(t,\hat{y}) = \int (r^4 dt\ d\hat{y})\ \tilde{L}_0(y)\ . \tag{18.17}$$

Therefore, the following measure is (quasi) invariant

$$(dy) = r^4 dt\ d\hat{y}\ . \tag{18.18}$$

This is not strictly speaking a measure on the projective cone, but it can be used to integrate scalar fields of degree -4 since then the r-dependence cancels out.

Now consider the SO(4,2) transformation (18.7). It induces a transformation on r,t and $\hat{y}$ that we denote as follows:

$$r \to r_\Lambda , \quad t \to t_\Lambda , \quad \hat{y} \to \hat{y}_\Lambda . \tag{18.19}$$

Since (dy) is invariant, the Jacobian is

$$J_\Lambda = \left| \frac{\partial t_\Lambda \, \partial \hat{y}_\Lambda}{\partial t \, \partial \hat{y}} \right| = (r/r_\Lambda)^4 . \tag{18.20}$$

If $\tilde{L}(y) = r^{-4} L(t,\hat{y})$, then the transformation law for a scalar field,

$$T_{\Lambda^{-1}} \tilde{L}(y) = \tilde{L}(\Lambda y)$$

becomes

$$T_{\Lambda^{-1}} L(t,\hat{y}) = r^4 \, T_{\Lambda^{-1}} \tilde{L}(y) = (r/r_\Lambda)^4 \, L(t_\Lambda, \hat{y}_\Lambda) . \tag{18.21}$$

This is a multiplier representation, with multiplier $(r/r_\Lambda)^4$. Invariance of the integral is proved by

$$T_{\Lambda^{-1}} \int dt \, d\hat{y} \, L(t,\hat{y}) = \int dt \, d\hat{y} \, (r/r_\Lambda)^4 \, L(t_\Lambda, \hat{y}_\Lambda)$$

$$= \int dt_\Lambda \, d\hat{y}_\Lambda \, L(t_\Lambda, \hat{y}) = \int dt \, d\hat{y} \, L(t,\hat{y}) . \tag{18.22}$$

The field $\phi(t,\hat{y}) = r\tilde{\phi}(y)$ also transforms by a multiplier representation,

$$T_{\Lambda^{-1}} \phi(t,\hat{y}) = (r/r_\Lambda) \, \phi(t_\Lambda, \hat{y}_\Lambda) . \tag{18.23}$$

The conventional Minkowski field $\hat{\phi}$ transforms in the same way, except that t,$\hat{y}$ are replaced by x_μ and r by y_4+y_5.

Finally, with this preparation we can see that the Lagrangian (18.12) is invariant and agrees with (18.1). Note especially Eq. (18.18) and the fact that the expression in the bracket in (18.12) has degree -4.

19. Quasi-invariant wave operator on U(2).

Dirac's projective cone is equivalent, as a homogeneous space for SO(4,2), to U(2) [up to a factor Z_2]. The correspondence is given by

$$U(2) \ni u = \vec{y}\cdot\vec{\sigma}/y_+ , \tag{19.1}$$

where

$$\vec{y}\cdot\vec{\sigma} = \sum_{i=1}^{4} y_i\sigma_i , \quad y_+ = y_5 + iy_0 = r\, e^{it} ,$$

$$y_i = r\hat{y}_i , \quad \sigma_4 = i , \quad \sigma_1,\sigma_2,\sigma_3 = \text{Pauli matrices} .$$

It is possible to parameterize U(2) by $0 \leq t < 2\pi$ and the unit vector $\hat{y} \in S_3$, except that $(t,\hat{y})$ must be identified with $(t+\pi, -\hat{y})$ for $0 \leq t < \pi$. The scalar field $\bar{\phi}$ of degree -1 on the cone corresponds to a two-valued function on U(2),

$$\phi(u) = \phi(t,\hat{y}) = r\bar{\phi}(y) . \tag{19.2}$$

This function of $t,\hat{y}$ was already introduced in Section 18, and its transformation law is (18.23). We shall rewrite the wave operator in terms of $\phi(u)$ and study its transformation properties.

When $r^2\,\partial_y^{\,2}$ acts on $\bar{\phi}$ it is

$$r^2\,\partial_y^{\,2} = \left(Y\frac{\partial}{\partial Y}\right)^2 - H^2 + Q_{so4} - r\frac{\partial}{\partial r}\left(r\frac{\partial}{\partial r} + 1\right) .$$

Here $H = L_{50} = i(y_5\partial_0 - y_0\partial_5)$ is the generator of the center of the compact subgroup, as before, and Q_{so4} is the Casimir operator of so(4):

$$Q_{so4} = \frac{1}{2}\sum_{i,j=1}^{4} L_{ij}^2 , \quad L_{ij} = i(y_i\partial_j - y_j\partial_i) .$$

Its spectrum is $\{\ell(\ell+2);\ \ell = 0,1,...\}$. Since the degree of ϕ is -1,

$$r^2\,\partial_y^{\,2} \to (\ell+1)^2 - E^2 \tag{19.3}$$

on each common eigenspace of Q_{so4} and H (eigenvalue E). This form is actually representation independent. The free wave equation $\partial_y^{\,2}\,\tilde{\phi}(y) = 0$ says that each eigenspace of H is an eigenspace of Q_{so4}, with $|E| = \ell+1$. Of course, more is true, for each eigenspace of H contains exactly one irreducible representation of so(4). This information is not directly expressed by the wave equation, but it may be deduced from it if we use the above expression for L_{ij} as an operator on functions on S_3.

The right hand side of (19.3) is equally applicable to $\phi(u)$, and we define $\Box$ by

$$\Box\phi(u) = \frac{1}{4}\,[(\ell+1)^2 - H^2]\,\phi(u) . \tag{19.4}$$

An elegant expression for $\Box$ will be obtained below. First it should be stressed, however, that though the free wave equation $\Box\phi(u) = 0$ is invariant, the operator $\Box$ is not. There are two ways of looking at this phenomenon.

In the first place, (19.3) is a non-central element of the enveloping algebra of so(4,2). With the usual notation it is

$$r^2\,\partial_y^{\,2} = 1 + \frac{1}{2}\sum_{i,j=1}^{4} L_{ij}^2 - L_{50}^2 . \tag{19.5}$$

This is related to a tensor operator,

$$J_{\alpha\beta} = \frac{1}{2}\delta^{\gamma\delta}(L_{\alpha\gamma}L_{\beta\delta} + L_{\beta\gamma}L_{\alpha\delta}) + \delta_{\alpha\beta}$$

$$\sum_{\alpha=0}^{5} J_{\alpha\alpha} = -2r^2\,\partial^2 .$$

The positive energy solutions of the free wave equation span a space of functions that carries an irreducible representation of SO(4,2). This is a very singular representation, characterized by a large ideal in the enveloping algebra. In particular, the tensor operator ($J_{\alpha\beta}$) vanishes on the solution space:

$$J_{\alpha\beta}\,\phi = 0\,, \quad \text{for free fields .}$$

This is, in fact, easy to show, for when we evaluate the action of $J_{\alpha\beta}$ on scalar fields of degree -1 on the cone, then we find

$$J_{\alpha\beta}\,\tilde{\phi}(y) = -y_\alpha y_\beta \partial_y^{\,2}\,\tilde{\phi}(y) .$$

On the other hand, the operator $\partial_y^{\,2}$ is certainly invariant, and the great virtue of Dirac's formalism is precisely the manifest invariance of the wave operator. The trouble is that $\partial_y^{\,2}$ is not an operator acting in the space of the representation. This space consists of fields $\tilde{\phi}$ of degree -1, but $\partial_y^{\,2}\,\tilde{\phi}$ is of degree -3 and lies in another space. In other words, $\partial_y^{\,2}$ is actually an intertwining operator. The factor r^2 is required to bring the degree back up to -1, and $r^2\,\partial_y^{\,2}$ is only quasi-invariant. We now determine its transformation properties.

We derived (18.21) from $L(t,\hat{y}) = r^4\,\tilde{L}(y)$, and (18.23) from $\phi(t,\hat{y}) = r\tilde{\phi}(y)$. In general, each power of r contributes one power of (r/r_Λ) to the multiplier. Since $\phi(u) = r\tilde{\phi}(y)$ and $4\Box\phi(u) = r^3\,\partial_y^{\,2}\,\tilde{\phi}(y)$ we get

$$T_{\Lambda^{-1}}\,\phi(u) = \frac{r}{r_\Lambda}\,\phi(u_\Lambda)\,,\quad T_{\Lambda^{-1}}\,\Box\phi(u) = \left(\frac{r}{r_\Lambda}\right)^3 \Box\phi(u_\Lambda)\,. \tag{19.6}$$

The fact that the multiplier is different shows that $\Box$ intertwines between two different representations.

We record here an especially elegant expression for the wave operator. Using the identity

$$\sum_{i=1}^{4} (\sigma_i)_{ab}\,(\sigma_i)_{cd} = -2\varepsilon_{ac}\varepsilon_{bd}\,,\quad \varepsilon_{ab} = (i\sigma_2)_{ab}\,,$$

we find that, if $\phi_+(u)$ is a polynomial in u,

$$y_+^{\,2}\,\partial_y^{\,2}\,\phi_+(u) = 4\,(\det \partial_u)\,\phi_+(u)\,. \tag{19.7}$$

The description of our scalar field module in terms of functions holomorphic in D_o will be discussed in the next section. For such functions the free wave equation is

$$\Box\phi_+(u) = 0\,, \tag{19.8}$$

$$\Box = \det \partial_u = \det(\partial/\partial u_{ab})\,. \tag{19.9}$$

20. The conformal tube.

The solution space S of the free wave equation is spanned by the energy eigenfunctions

$$\phi_{E\ell m}(u) = e^{iEt}\,Y_{\ell m}(\hat{y})\,,\quad E = \pm(\ell+1)\,,\quad \ell = 0,1,\ldots, \tag{20.1}$$

where the $Y_{\ell m}(\hat{y})$ are 4-dimensional spherical harmonics, an orthonormal basis for $L^2(S_3,d\hat{y})$. We shall now, until further notice, confine ourselves to the positive energy subspace S_+, then we can represent $\phi(u)$ as

$$\phi(u) = e^{-it}\,\phi_+(u)\ , \tag{20.2}$$

where ϕ_+ is a polynomial in the matrix elements of u. We can therefore look at S_+ as the space of boundary values of a space H_+ of functions that are holomorphic on the domain

$$D_o :\ 1 - uu^\dagger > 0\ .$$

Let us now study this domain, and its boundary.

The center of u(2,2) consists of matrices that are multiples of the unit matrix; Eq. (14.10) shows that it acts trivially, and only $\mathscr{C}$ = $SU(2,2)\,/\,Z_2 \otimes Z_2$ acts effectively. This group acts transitively on D_o; if we include the boundary then we find three orbits:

$$\text{(i) } X_o' = \text{space time, dimension 4, } 1 - uu^\dagger = 0\ , \tag{20.3}$$

$$\text{(ii) } X_1',\ \text{dimension 7, } 1 - uu^\dagger \geq 0,\ \det(1-uu^\dagger) = 0\ , \tag{20.4}$$

$$\text{(iii) } X_2' = D_o,\ \text{dimension 8, } 1 - uu^\dagger > 0\ . \tag{20.5}$$

Compare the orbits of $U^\gamma(4)$ in X, Eqs. (15.8)-(15.10). The space X_o' is canonically identified with X_o, and the prime may be omitted. The spaces X_1' and X_2' are related to X_1 and X_2 by projections that essentially amount to ignoring the variables $v,\bar{v}$.

We shall find expressions for these orbits in Dirac's notation. Extend (19.1) to the complex domain and define $y_- = y_5 - iy_o$ so that y^2 remains zero:

$$u = \vec{y}\cdot\vec{\sigma}/y_+ \,, \quad y_+y_- - \vec{y}^2 = y^2 = 0 \,, \quad y_\pm = y_5 \pm iy_o \,. \tag{20.6}$$

Let ξ and η be the real and imaginary parts of y:

$$y = \xi + i\eta \,, \quad y^2 = \xi^2 - \eta^2 + 2i\xi\cdot\eta = 0 \,.$$

The group SO(4,2) acts on y by pseudo-orthogonal matrices,

$$y \to \Lambda y = \Lambda\xi + i\Lambda\eta \,.$$

We consider the orbits of this action in the complex cone. Every orbit contains a point such that $\eta_i = 0$, $i = 1,2,3$. Then u is fully reducible and the eigenvalues are

$$|\langle u\rangle| = (x+iy_4)/y_+ \,, \quad x^2 = \xi_1{}^2 + \xi_2{}^2 + \xi_3{}^2 \,,$$

$$|\langle u\rangle|^2 = [(x-\eta_4)^2 + \xi_4{}^2]/[(\xi_5-\eta_o)^2 + (\xi_o+\eta_5)^2] \,.$$

Suppose first that $\eta^2 = \xi^2 < 0$, then there is a point on the orbit such that $\eta_\alpha = 0$ for $\alpha \neq 4$, and $\xi_4 = 0$. Here

$$|\langle u\rangle|^2 = (x-\eta_4)/(x+\eta_4) \,,$$

and it is seen that one eigenvalue of u lies inside the unit circle while the other lies outside. Next, suppose that $\xi^2 = \eta^2 = 0$. Then there is a point on the orbit at which

$$(\eta_\alpha) = (0,0,0,0,1,1) \,, \quad (\xi_\alpha) = (a,a,0,0,b,b) \,,$$

$\alpha = 0,\ldots,5$. In this case $x^2 = a^2$ and

$$|\langle u\rangle|^2 = [(x-1)^2 + b^2]/[b^2 + (a+1)^2] .$$

We see that one eigenvalue of u has modulus 1, and the other one lies inside the unit circle if and only if a is positive.

Let us digress for a moment, and consider ξ and η as two points on the real cone. The relationship between ξ and η that expresses the fact that $\xi + i\eta$ maps into D_o or its boundary is of course invariant; in fact, it is precisely the relationship that defines the causal structure on the cone: ξ is in the invariant future of η. This can be expressed invariantly by the inequality $\xi_o\eta_5 - \xi_5\eta_o > 0$. We shall write, when $\xi^2 = \eta^2 = 0$, and $y = \xi + i\eta$,

$$y > 0 \Longleftrightarrow \xi > \eta \Longleftrightarrow \xi_o\eta_5 - \xi_5\eta_o > 0 . \tag{20.7}$$

We emphasize that this ordering relation is SO(4,2) invariant. It is also a projective invariant, if $y > 0$ then also $\lambda y > 0$ for any complex scalar λ. This domain maps onto X_1'. The limit points, where the above inequality becomes an equality, consists of points where $y = \lambda\xi$, with λ a complex scalar; they map onto X_o'.

Finally, suppose that $\xi^2 = \eta^2 > 0$. Then a similar calculation shows that the domain that maps onto D_o consists of the points where $y > 0$, in the sense of (20.7), the definition of this causal relationship being extended to the domain $\xi^2 = \eta^2 \geq 0$.

We now exclude the points where y_4+y_5 vanishes (Minkowski infinity), and go over to the Minkowski notation. Since $y \simeq \lambda y$ for any complex $\lambda \neq 0$, we may put $y_4+y_5 = 1$ and

$$x^\mu + y^\mu = \xi^\mu + i\eta^\mu , \quad \mu = 0,1,2,3 .$$

Since $y^2 = 0$, we have

$$y_4 + y_5 = 1 \, , \; y_4 - y_5 = x^2 \equiv x_o^{\,2} - \vec{x}^2 \, ,$$

with arrows adorning 3-vectors in this context. So far, there are no conditions on the complex vector x. The condition $\xi^2 = \eta^2 > 0$ (on the six-vectors) translates amazingly to the four-vector condition

$$\eta^2 = \eta_o^{\,2} - \vec{\eta}^2 > 0 \, . \tag{20.8}$$

The final condition (20.7) also becomes remarkably simple in this case; so long as (20.8) holds it reduces to

$$\eta_o < 0 \, . \tag{20.9}$$

The domain given by (20.8) and (20.9), with $\eta_\mu = \mathrm{Im}\, x_\mu$, is the forward tube of relativistic field theories on Minkowski space. We have thus found that this domain is precisely the bounded symmetric domain D_o.

21. Hilbert space of holomorphic functions.

We return to the space S_+ of positive energy solutions of the free wave equation, being the span of the functions (20.1). We have the orthonormality property (for $E = \ell+1$, $E' = \ell+1$)

$$\int d\hat{y} \; \phi^*_{E\ell m}(t,\hat{y}) \; \phi_{E'\ell' m'}(t,\hat{y}) = \delta_{\ell\ell'} \delta_{mm'} \, ,$$

however, the $L^2(S_3, d\hat{y})$ norm is not invariant. Neither is the $L^2(S_1 \otimes S_3, dtd\hat{y})$ norm, since the measure $dtd\hat{y}$ is only quasi-invariant. However, from the invariance of the Lagrangian it follows in standard fashion that the inner product

$$(\phi,\phi') = \int d\hat{y}\ \phi^*(t,\hat{y})\ \frac{i}{2}\ \overleftrightarrow{\partial}_t\ \phi(t,\hat{y}) \tag{21.1}$$

is independent of t and in fact invariant under the conformal group. It is not definite, but it becomes positive definite when we confine ourselves to the positive energy solutions of the wave equation. Because (21.1) is the type of invariant inner product that is used in the formulation of Cauchy initial value problems, we shall refer to it as the initial data norm. We use it to turn S_+ into a Hilbert space. In order that our basis functions be orthonormal with respect to (21.1) we must normalize them differently, and we replace (20.1) by

$$\phi_{E\ell m}(t,\hat{y}) = (\ell+1)^{-1/2}\ e^{-iEt}\ Y_{\ell m}(\hat{y})\ ,\quad E = \ell+1\ . \tag{21.2}$$

We shall now see if (21.1) can be replaced by an integral over D_o.

An invariant measure on D_o is found as follows. We first verify that the line element

$$ds^2 = \mathrm{tr}\{(1-u^\dagger u)^{-1}\ du^\dagger\ (1-uu^\dagger)^{-1}\ du\} \tag{21.3}$$

is invariant under the action $u \to g*u$ of SU(2,2). This is the Bergman metric. Therefore, the two-form

$$\Omega = \mathrm{tr}\{(1-u^\dagger u)^{-1}\ du^\dagger \wedge (1-uu^\dagger)^{-1}\ du\} \tag{21.4}$$

is also invariant, and so is the volume element

$$(du) = \mathrm{const.}\ \Omega \wedge \Omega \wedge \Omega \wedge \Omega\ . \tag{21.5}$$

Let us write

$$du du^\dagger = \prod_{i,j} dx_{ij} \wedge dy_{ij} , \quad u_{ij} = x_{ij} + iy_{ij} ,$$

for the ordinary Lebesque measure, and choose the constant so that

$$(du) = [\det(1-uu^\dagger)]^{-4} \, du du^\dagger . \tag{21.6}$$

This is, up to a constant factor, the only invariant measure on D_o.

Recall that the functions

$$\phi_+(u) = e^{it} \phi(t,\hat{y}) = y_+ \tilde{\phi}(y) \tag{21.7}$$

are boundary values of functions holomorphic in D_o. Their transformation properties are determined by those of $\tilde{\phi}$,

$$T_{\Lambda^{-1}} \phi_+(u) = \frac{y_+}{(\Lambda y)_+} \phi_+(u_\Lambda) . \tag{21.8}$$

This is exactly the same as the formula (14.13):

$$g^{-1} \phi_+(u) = [\det(ru+s)]^{-1} \phi_+(g*u) , \tag{21.9}$$

$$g*u = (pu+q)(ru+s)^{-1} , \quad \begin{pmatrix} p & q \\ r & s \end{pmatrix}^* = g \in SU(2,2) . \tag{21.10}$$

From this last formula one obtains

$$|\det(ru+s)|^2 \, g * \det(1-uu^\dagger) = \det(1-uu^\dagger) ,$$

and thus

$$\int_{D_o} (du)\,[\det(1-uu^\dagger)]^{-N}\, g^{-1}\,[\phi_+(u)^*\,\phi_+'(u)]$$

$$= \int_{D_o} (dg{*}u)\,[\det(1-uu^\dagger)]^{-N}\,|\det(ru+s)|^{-2N-2}\,\phi_+(g{*}u)^*\,\phi_+'(g{*}u)$$

$$= \int_{D_o} (du)\,[\det(1-uu^\dagger)]^{-N}\,\phi_+(u)^*\,\phi_+'(u)\ ;$$

the last equality being true provided N = -1. Therefore, the sesquilinear form

$$(\phi_+,\phi_+') = \int (du)\,[\det(1-uu^\dagger)]\,\phi_+(u)^*\,\phi_+'(u)$$

$$= \int_{D_o} du du^\dagger\,[\det(1-uu^\dagger)]^{-3}\,\phi_+(u)^*\,\phi_+'(u) \qquad (21.11)$$

is invariant under the action (21.9).

Thus it seems that we may have succeeded in expressing the inner product as an integral over D_o. We only have to fear the following two possibilities: the integral (21.11) may diverge, or it may vanish on our module.

As a preliminary to regularizing the integral, and in order to get a wider perspective on this problem, let us consider the generalization of (21.9) given by

$$g^{-1}\phi_+(u) = [\det(ru+s)]^N\,\phi_+(g{*}u)\ . \qquad (21.12)$$

We shall avoid the issue of making sense of this expression for non-integral N, because only integral values of N will be strictly relevant to our problem. We restrict ourselves to the neighborhood of g = 1, at which point the multiplier takes the value 1. The action of the Lie algebra on the polynomials is a highest weight module that we denote $\mathcal{F}(E_o)$, where $E_o = -N$ is the lowest energy and the minimal so(4)-type is trivial. For this action the invariant generalization of (21.11) is

$$(\phi_+,\phi_+') = \int_{D_o} dudu^\dagger\, [\det(1-uu^\dagger)]^{E_o-4}\, \phi_+(u)^*\, \phi_+'(u)\,.$$

Note that the determinant is positive on D_o so that no problem arises when E_o is non-integral, provided only that the action (21.12) is so defined that

$$g^{-1}[\phi_+(u)^*\, \phi_+'(u)] = |\det(ru+s)|^{2N}\, \phi_+(g*u)^*\, \phi_+'(g*u)\,. \qquad (21.13)$$

For $E_o > 3$ the integral converges. In particular,

$$\int_{D_o} dudu^\dagger\, [\det(1-uu^\dagger)]^{E_o-4} = \pi^4/(E_o-3)(E_o-2)^2(E_o-1)\,. \qquad (21.14)$$

The integral diverges for $E_o = 1$. It can be regularized by multiplication by (E_o-1) and passing to the limit. In this case the integral concentrates on the boundary, in the sense that the contribution of any closed interior subset of D_o vanishes in the limit. In the next section we shall obtain the regularization of (21.14) as an integral over the characteristic manifold X_o.

22. Invariant bilinear functionals.

Recall that X_o, the space of unitary, 2-dimensional matrices, is the characteristic manifold of D_o. Let us now study invariant bilinear, or more properly, sesquilinear functionals over X_o.

The evaluation of the invariant volume element (21.6) via the line element (21.3) can be repeated to show that

$$dudv\, |\det(1-uv^\dagger)|^{-4}\,, \quad u,v \in X_o\,, \quad du = dtd\hat{y}\,, \qquad (22.1)$$

is invariant. Also, if g acts on u and on v, we have

$$\det(ru+s)\det(rv+s)^{\dagger}\; g{*}\det(1-uv^{\dagger}) = \det(1-uv^{\dagger}) \,. \tag{22.2}$$

Therefore, if g acts on ϕ_+ as in (21.12),

$$g^{-1}\phi_+(u) = [\det(ru+s)]^{-E_o}\,\phi_+(g{*}u) \,,$$

then the following integral is superficially invariant:

$$\int_{X_o\times X_o} du dv\, |\det(1-uv^{\dagger})|^{-4}\, [\det(1-vu^{\dagger})]^{E_o}\, \phi_+(u)^{*}\, \phi_+'(v) \,. \tag{22.3}$$

This formula is meaningless as it stands, unless $E_o = 4,5,\ldots$, but perhaps the integral can be interpreted in the sense of its regularization, even at the critical points $E_o = 1,2,3$. As a preliminary, we shall return to Dirac's notation to get some inspiration.

Integrals over the cone are invariant if the integrand is a scalar field of degree -4. Examples are

$$\int (dy)\, \tilde\phi^{*}(y)\, \tilde\psi(y) \,, \quad \text{deg. } \tilde\phi + \text{deg. } \tilde\psi = -4 \,, \tag{22.4}$$

$$\int (dy)\, \tilde\phi^{*}(y)\, \partial_y^{\,2}\, \tilde\phi(y) \,, \quad \text{deg } \tilde\phi = -1 \,, \tag{22.5}$$

$$\int (dy)(dy')(y{\cdot}y')^{-\nu}\, \tilde\phi^{*}(y)\, \tilde\psi(y') \,, \quad \text{deg. } \tilde\phi = \text{deg. } \tilde\psi = \nu-4 \,, \tag{22.6}$$

$$\tilde\phi(y) = \int (dy')(y{\cdot}y')^{-\nu}\, \tilde\psi(y') \,, \quad \text{deg. } \tilde\phi = -\nu \,, \ \text{deg. } \tilde\psi = \nu-4 \,. \tag{22.7}$$

We suspect that (22.6) is related to (22.1).

To make sense of the above, let us begin by defining the generalized function $(y{\cdot}y')^{-\nu}$. Suppose first that ν is not an integer.

Interpreting $(y{\cdot}y')^{-\nu} = x^{-\nu}$ as a function of a single variable $x = y{\cdot}y'$ we have the following possibilities: $(x-io)^{-\nu}$, $(x+io)^{-\nu}$, and linear combinations of these two. Since the functions $\bar{\phi},\tilde{\psi}$ being considered, for the time being at least, have positive energies, it is sensible to define $(y{\cdot}y')^{-\nu}$ as a Fourier series with positive frequencies:

$$(2y{\cdot}y')^{-\nu} = (rr')^{-\nu} \sum_{n=0}^{\infty} e^{-i\tau(n+\nu)} C_n^{\nu}(\hat{y}{\cdot}\hat{y}') , \quad \tau = t-t' . \tag{22.8}$$

This is the same as defining $(2y{\cdot}y')^{-\nu}$ as the boundary values of a function analytic in Im $\tau > 0$; that is, in the tube discussed in Section 20. The C_n^{ν} are Gegenbauer polynomials. Now (22.8) is perfectly well defined for any real ν, and that makes (22.6) well defined as well. On the other hand, the apparent invariance of these integrals is an illusion unless ν is integer, for only then are the functions $\exp[-i\tau(n+\nu)]$ one-valued.

To translate this to the matrix notation we set

$$u = \hat{y}{\cdot}\hat{\sigma}/y_+ , \quad v = \hat{y}'{\cdot}\hat{\sigma}/y_+'$$

and find that

$$\det(1-uv^{\dagger}) = (y_+y_-')^{-1} (2y{\cdot}y') = e^{-i\tau} (rr')^{-1} (2y{\cdot}y') . \tag{22.9}$$

Of course, powers of this determinant occurring in (22.3) should be interpreted as limiting values from D_0, which is precisely what (22.8) implies if we interpret the determinants in (22.3) as follows:

$$|\det(1-uv^{\dagger})|^{-4} [\det(1-vu^{\dagger})]^{E_0} = (\det uv^{\dagger})^{2-E_0} [\det(1-uv^{\dagger})]^{E_0-4} , \tag{22.10}$$

and

$$[\det(1-uv^\dagger)]^{-\nu} = (rr')^\nu e^{i\nu\tau} (2y\cdot y')^{-\nu}$$

$$= \sum_{n=0}^{\infty} e^{-in\tau} C_n^{\nu}(\hat{y}\cdot\hat{y}') . \qquad (22.11)$$

We are thus assured that (22.3) has a meaning; next, let us find out what is is.

To this end we expand the Gegenbauer polynomials C_n^{ν} in terms of the SO(4) spherical functions C_n^{1}:

$$C_n^{\nu}(\hat{y}\cdot\hat{y}') = \sum_{p=0}^{[n/2]} \frac{(\nu-1)\nu \ ... \ (\nu+p-2) \ \nu(\nu+1) \ ... \ (\nu+n-p-1)}{p! \ (n-p+1)!}$$

$$\times (n+1-2p) \, C^1_{n-2p}(\hat{y}\cdot\hat{y}') . \qquad (22.12)$$

To read the information contained in this formula recall that

$$\frac{\ell+1}{2\pi^2} C_\ell^{1}(\hat{y}\cdot\hat{y}') = \sum_m Y_{\ell m}(\hat{y}) \, Y_{\ell m}(\hat{y}')^* \qquad (22.13)$$

is the integral kernel for the projection operator to an irreducible SO(4) module (reproducing kernel). [As an SU(2) $\otimes$ SU(2) module it is the representation with indices $\ell_1 = \ell_2 = \ell/2$ and dimension $(\ell+1)^2$.

Let us discuss the full operator kernel $(y\cdot y')^{-\nu}$ defined by (22.8), rather than (22.11). To each value of n in (22.8) there corresponds a frequency component of $(y\cdot y')^{-\nu}$ of energy $n+\nu$. If $\nu \neq 1,0,-1,...$, then none of the coefficients in (22.12) vanish. It is then evident that the K-types that appear in the expansion of $(y\cdot y')^{-\nu}$ are precisely the same (including multiplicities) as those that make up the irreducible, highest weight su(2,2) module $D(\nu,0,0)$ with minimal energy ν and trivial lowest so(4) type. If $\nu > 1$, then all the coefficients in (22.12) are positive, which shows that $D(\nu,0,0)$ is unitary (more precisely, unitarizable) in this case. If $\nu < 1$, then some coefficients are negative and $D(\nu,0,0)$ is certainly not unitary in this case. When $\nu = 1$, only the term $p = 0$ survives, and the reduced K-structure of $D(1,0,0)$ is obtained. For $\nu = 0,-1,\ldots$ the number of terms surviving is 2,3,... (less for the lowest values of n). Whatever the value of ν, (22.8) is always a reproducing kernel for the irreducible representation $D(\nu,0,0)$.

It would seem, at first, that the case $\nu = 4 - E_0 = 3$ is the only case of direct interest to us, but it will soon become clear that we need to study some other cases as well.

$\underline{\nu = 3}$. We saw that $(y\cdot y')^{-3}$, as defined by (22.8), is a reproducing kernel for $D(3,0,0)$. The K-spectrum of this representation has no overlap with that of $D(1,0,0)$. Consider the operator defined by (22.7), when $\nu = 3$:

$$\tilde{\phi}(y) = \int (dy')(2y\cdot y')^{-3}\, \tilde{\psi}(y')\ , \quad \deg.\ \tilde{\psi} = -1\ . \tag{22.14}$$

Evidently, $\tilde{\phi}$ must vanish if $\tilde{\psi}$ belongs to our $D(1,0,0)$ module. But $D(1,0,0)$ is highly degenerate, and there are functions of degree -1 that it makes no use of. To be precise, consider the action of so(4,2) on the K-finite vectors in the space of functions of degree -1 and positive energy. This action is indecomposable, of the type

$$D(3,0,0) \rightarrow D(1,0,0)\ . \tag{22.15}$$

[Notation: If A,B are representations, then $A \rightarrow B$ is a non-decomposable representation in which B acts in an invariant subspace and A acts in a quotient space.] The operator (22.14) acts on this space of functions of degree -1 and produces functions of degree -3. Its kernel is our $D(1,0,0)$ module. It intertwines between (22.15) and the irreducible D(3,0,0) module that consists of functions of degree -3. We translate this to the realization by functions holomorphic in D_o. To fields of degree -1 correspond the action of SU(2,2) with multiplier $[\det(ru+s)]^{-1}$ as in (21.9). When this action is applied to the space spanned by the polynomials in u, then it is equivalent to (22.15). The operator (22.14) takes the form

$$\phi_+(u) = \int_{X_o} dv\,(-\det v^\dagger)\,[\det(1-uv^\dagger)]^{-3}\,\psi_+(v)\,. \tag{22.16}$$

It is easy to check that, if ψ_+ takes multiplier $[\det(ru+s)]^{-1}$, then ϕ_+ takes multiplier $[\det(ru+s)]^{-3}$. All this suggests that the operator (22.14) or (22.16) is closely related to the wave operator that we investigated in Section 19. We shall confirm this presently.

As far as the interpretation of (22.3) as an inner product is concerned, the conclusion is that this formula defines an inner product for D(3,0,0), realized on the space of functions on X_o that extend to holomorphic functions on D_o, modulo the invariant subspace that carries our massless representation D(1,0,0).

23. Cauchy kernel and Lagrangian.

We continue investigating the content of (22.3) and (22.7) in some interesting cases.

$\nu = 2$. Eq. (22.11) becomes

$$C_n^2(\hat{y}\cdot\hat{y}') = \sum_\ell (\ell+1)\,C_\ell^1(\hat{y}\cdot\hat{y}') = 2\pi^2 \sum_{\ell m} Y_{\ell m}(\hat{y})\,Y_{\ell m}(\hat{y}')^* \tag{23.1}$$

The sum is over the set $\ell = n, n-2, \ldots, 1$ or 0. Consequently

$$[\det(1-uv^\dagger)]^{-2} = 2\pi^2 \sum_{n\ell m} [e^{-int} Y_{\ell m}(\hat{y})][e^{-int'} Y_{\ell m}(\hat{y}')]^*. \tag{23.2}$$

This is the Cauchy kernel of the domain D_o; it has the property

$$\int_{X_o} dt'd\hat{y}' \, [\det(1-uv^\dagger)]^{-2} \, \psi_+(v) = 4\pi^3 \, \psi_+(u) \, , \quad u \in D_o \, , \tag{23.3}$$

for any function ψ_+, holomorphic in D_o and infinitely differentiable on the boundary. Similarly, if deg $\tilde{\psi}$ is -2,

$$\int (dy)(2y\cdot y')^{-2} \, \tilde{\psi}(y') = 4\pi^3 \, \tilde{\psi}(y) \, . \tag{23.4}$$

Now, let us return to the case $\nu = 3$.

When $\nu = 3$, Eq. (22.11) becomes

$$C_n^3(\hat{y}\cdot\hat{y}') = -\frac{1}{8}\sum_\ell [(\ell+1)^2 - E^2] \, (\ell+1) \, C_\ell^1(\hat{y}\cdot\hat{y}') \, , \quad E = n+3 \, . \tag{23.5}$$

The factor $(-1/8)[(\ell+1)^2 - E^2]$ that constitutes the difference between this function and (23.1) is just (-1/8) times the "Fourier" coefficient of the wave operator, Eq. (19.3). Consequently

$$\int (dy')(2y\cdot y')^{-3} \, \tilde{\psi}(y') = -\frac{1}{2}\pi^3 \, \partial_y^{\,2} \, \tilde{\psi}(y) \, . \tag{23.6}$$

In view of the relation (19.7),

$$y_+^{\,2} \, \partial_y^{\,2} = 4\Box \, ,$$

this is the same as

$$\int_{X_o} dv \det v^\dagger \, [\det(1-uv^\dagger)]^{-3} \, \psi_+(v) = 2\pi^3 \, \Box\psi_+(u) \, . \tag{23.7}$$

We can also verify this result directly, as follows.

First, it is easy to check that

$$\Box[\det(1-uv^\dagger)]^{-2} = 2 \det v^\dagger \, [\det(1-uv^\dagger)]^{-3} . \tag{23.8}$$

Applying $\Box$ to (23.3) we get (23.7).

Let us summarize the results. The experience that we are gaining here will help us to work out the analogous problem for the oscillator representation of Sp(8).

First of all, we have learned that all our efforts to find a formula for the invariant inner product has led instead to the free action integral:

$$\mathcal{L} = -\frac{1}{2}\int (dy)\, \tilde{\psi}(y)^* \, \partial_y^{\,2} \, \tilde{\psi}(y) \tag{23.9}$$

$$= \pi^{-3} \int (dy)(dy')\, \tilde{\psi}(y)^* \, (2y\cdot y')^{-3} \, \tilde{\psi}(y') \tag{23.10}$$

$$= \pi^{-3} \int du dv \, \det uv^\dagger \, \phi_+(u)^* \, [\det(1-uv^\dagger)]^{-3} \, \phi_+(v) \tag{23.11}$$

$$= 2 \int du \, \det u \, \phi_+(u)^* \, \Box\phi_+(u) , \tag{23.12}$$

with $\Box = \det \partial_u$. It must be stressed that the double integrals are valid expressions only on certain domains of holomorphic or positive energy modules. Only the first and the last expressions are suitable for the formulation of dynamics.

Secondly, we have some interesting results concerning generalized functions on U(2). If we define the Dirac δ-function relatively to the measures (dy) or du = $dt d\hat{y}$, then

$$(2y\cdot y')^{-2} = 4\pi^3 \, \delta(y,y') \qquad ,$$

$$\det v^\dagger \, [1-uv^\dagger)]^{-2} = 4\pi^3 \, \delta(t-t') \, \delta(\hat{y},\hat{y}') . \tag{23.13}$$

Each δ-function has its own space of test functions, as may be inferred from the above. Also

$$(2y\cdot y')^{-3} = -\frac{1}{2}\pi^3 \partial_y{}^2 \delta(y,y')$$

$$\det v^\dagger [\det(1-uv^\dagger)]^{-3} = 2\pi^3 \Box\delta(t-t')\, \delta(\hat{y},\hat{y}') . \qquad (23.14)$$

As far as generalized functions on D_o are concerned, we are led by these results to formulate the following.

Conjecture #1. The generalized functions on D_o,

$$\lim_{\nu\to 2} (\nu-2)^2 [\det(1-uu^\dagger)]^{-\nu} \quad ,$$

$$\lim_{\nu\to 2} (\nu-2)^2 [\det(1-uu^\dagger)]^{-\nu-1} \quad ,$$

are concentrated on the characteristic manifold.

Conjecture #2.

$$\lim_{\nu\to 2} (\nu-2)^2 \int_{D_o} du du^\dagger [\det(1-uu^\dagger)]^{-\nu} |\phi_+(u)|^2$$

$$= (-\pi/4) \int_{X_o} du\, |\phi_+(u)|^2 \quad , \qquad (23.15)$$

$$\lim_{\nu\to 2} (\nu-2)^2 \int_{D_o} du du^\dagger [\det(1-uu^\dagger)]^{-\nu-1} |\phi_+(u)|^2$$

$$= (\pi/8) \int_{X_o} du \det u\, \phi_+(u)^* \Box\phi_+(u) . \quad (23.16)$$

VII. osp(8) Field Theory--A Beginning

This Part brings us to the edge of the unknown. Applying what was learned in Part VI, we are quickly led to suspect that osp(8) invariant wave equations must be of very high order. This is not the final word on the subject, however. The origin of the difficulty is the very high degree of degeneracy of the relevant representation. This is not without precedents in field theories. In fact, one encounters exactly the same problem in conformal field theories and in extended supersymmetric field theories. One may hope that the massive effort that is now being directed towards a complete formulation of such theories may lead to new insight into the osp(8) problem as well. Perhaps the present effort to put the problem into focus may also prove helpful in the development of extended supersymmetric theories.

24. Intertwining operators.

We here begin to study wave equations and related problems for the free field that carries the even oscillator representation. There is close analogy with the scalar field theory of Sections 18-23, and we shall use much of the same notation. From the standpoint of logical development, however, Sections 13-23 should be considered as a digression; we are now picking up the main thread where it was dropped at the end of Section 12. Thus $\phi(w)$ is the value at w of a function holomorphic in D, or its boundary value on X. The action of SP(2n) was given by (11.14) and (11.15), with $\alpha = -1/2$. As in Sections 21-23, we consider the general case of real α; then the minimal energy is $E_o = -2\alpha$ and (11.14) becomes

$$g^{-1}\,\phi(w) = [\det M(g,w)]^{-E_o/2}\,\phi(g{*}w)\,. \tag{24.1}$$

The matrix M(g,w) was given by (11.4).

This action, restricted to the Lie algebra, and applied to the space of polynomials in w, is a realization of the lowest weight module $\mathcal{F}(\lambda) = \mathcal{F}(E_0,\bar{\lambda})$ with lowest energy E_0 and $\bar{\lambda} = 0$. The lowest weight is

$$\lambda = (\lambda_1,\lambda_1,\lambda_1,\lambda_1) = (E_0,\bar{\lambda}) ,$$

$$E_0 = 2\lambda_1 , \quad \bar{\lambda} = 0 \qquad ; \qquad (24.2)$$

compare Eqs. (5.8) and (5.11). Note that $\mathcal{F}(\lambda)$ consists of all the polynomials; its K-structure is independent of E_0. It is irreducible for $E_0 \geq 0$ except at the reduction points: $E_0 = 0,1,2,3$. The irreducible representation with lowest weight λ is denoted $\mathcal{D}(\lambda)$ or $\mathcal{D}(E_o,\bar{0})$. It is unitary if $E_0 > 3$ and at the reduction points.

The lowest weight λ is invariant under the Weyl reflection of $\lambda+\rho$ that commutes with su(n) when $\lambda_1 = 5/2$, $E_0 = 5$. We call this value of λ or λ_1 or E_0 the Weyl (symmetry) point. Fig. 1 summarizes the significance of the integral values of E_0.

Our goal is to find an invariant action, and we begin as in Section 21 by investigating invariant integrals over D. The invariant measure on D can be obtained in the same way:

$$(dw) = [\det(1-ww^*)]^{-5}\, dwdw^* .$$

The first factor is the Bergman kernel and

$$dwdw^* = \prod_{i\leq j} dx_{ij} \wedge dy_{ij} , \quad w_{ij} = x_{ij} + iy_{ij}$$

is the Lebesgue measure. Next we have

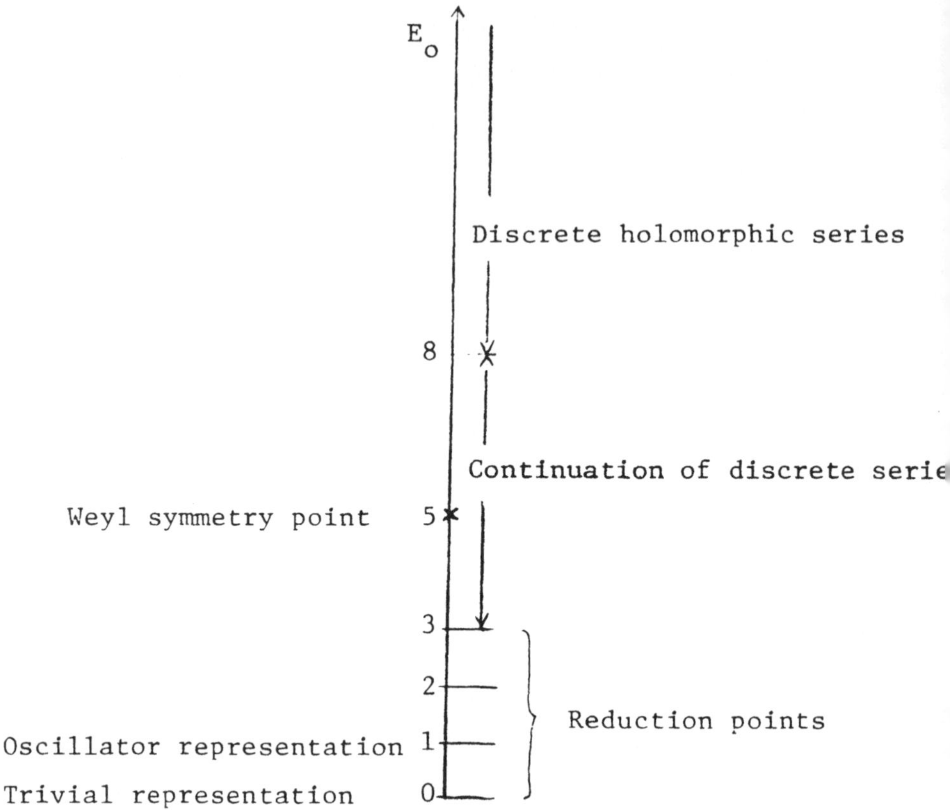

Fig. 1.

The lowest positive weights of sp(8/R) that are trivial on the semisimple part of the compact subalgebra. They are parameterized by the energy E_o, defined by Eq. (5.11) as the projection of the weight on the generator H of the center of the compact subalgebra, specialized to the case n = 4.

$$|\det M(g,w)|^2 \; g{*}\det(1-ww^*) = \det(1-ww^*)$$

and thus the invariant sesquilinear form

$$(\phi,\phi') = \int_D dwdw^* \, [\det(1-ww^*)]^{E_0/2 - 5} \, \phi(w)^* \, \phi'(w) \, . \qquad (24.3)$$

This is convergent for $E_0 > 8$. In particular,

$$\int_D dwdw^* \, [\det(1-ww^*)]^{E_0/2 - 5}$$

$$= 16\pi^{10}/(E_0-2)(E_0-3)(E_0-4)^2(E_0-5)^2(E_0-6)^2(E_0-7)(E_0-8) \, .$$

For $E_0 \leq 8$ we must regularize the integral.

It is known, especially from the work of Rossi and Vergne, that the appropriate measures for low, integral values of E_0 are concentrated on orbits of Sp(8) in the boundary of D. There are four such orbits, and our X is the one of lowest dimension. Instead of L^2 norms on all 4 orbits, or on D, we investigate general sesquilinear functionals on X.

The analogue of (22.3) is the formally invariant expression

$$\int_{X\times X} dwd\omega \, |\det(1-w\omega^*)|^{-5} \, [\det(1-\omega w^*)]^{E_0/2} \, \phi(w)^* \, \phi'(\omega) \qquad (24.4)$$

To make this meaningful we must interpret the kernel as the limit of a function of w, holomorphic in D:

$$|\det(1-w\omega^*)|^{-5} \, [\det(1-\omega w^*)]^{E_0/2}$$

$$= (\det w\omega^*)^{(5-E_0)/2} \, [\det(1-w\omega^*)]^{E_0/2 - 5} \, . \qquad (24.5)$$

In the special case $E_0 = 5$ this is the Cauchy kernel of X. If dw is normalized as in Hua, then

$$\int dw \, [\det(1-w\omega^*)]^{-5/2} \, \phi(\omega) = \frac{\pi}{3} (2\pi)^{10} \, \phi(w) \tag{24.6}$$

for ϕ the boundary value of a function holomorphic in D, and in this case (24.4) is proportional to

$$\int_X dw \, \phi(w)^* \, \phi'(w) \, .$$

Now it may be seen that $[\det(1-w\omega^*)]^{E_0/2 - 5}$ is the reproducing kernel for $\mathcal{D}(10-E_0,\bar{0})$. It is therefore clear that the formal invariance of (24.4) (checked without due regard for global properties) is deceptive unless $\mathcal{F}(E_0,\bar{0})$ contains $\mathcal{D}(10-E_0,\bar{0})$ as a subquotient. But this can happen only if E_0 is an odd integer less than or equal to 5.

We already discussed the case $E_0 = 5$. When $E_0 = 3$, then (24.4) reduces to

$$\int_{X\times X} dw d\omega \, \det(w\omega^*) \, [\det(1-w\omega^*)]^{-7/2} \, \phi(w)^* \, \phi'(\omega) \, . \tag{24.7}$$

Now, if

$$\Box = \det \partial_w \, ,$$

then

$$\Box[\det(1-w\omega^*)]^{-5/2} = \text{const.} \det(1-w\omega^*)^{-7/2} \det \omega^* \, .$$

In view of (24.6), this means that (24.7) is proportional to

$$\int_X dw \, \det w \, \phi(w)^* \, \Box\phi'(\omega) \, . \tag{24.8}$$

Furthermore, $E_o = 3$ or $\lambda = (3,\bar{0})$ is the first reduction point, and the highest weight of the invariant subquotient is $\lambda' = (7,\bar{0})$. The irreducible submodule $\mathcal{D}(3,\bar{0})$ is characterized by the wave equation $\Box\phi = 0$, and (24.8) is an invariant inner product for $\mathcal{D}(7,\bar{0})$. All this is in close analogy with our findings for the scalar field theory in Sections 22, 23.

We repeat that, as $E_o \to 3$, $\mathcal{F}(E_o,\bar{0})$ becomes reducible:

$$\mathcal{F}(3,\bar{0}) = \mathcal{D}(3,\bar{0}) \to \mathcal{D}(7,\bar{0}) \,, \tag{24.9}$$

and $\Box$ intertwines this to $\mathcal{F}(7,\bar{0})$. The case $E_o = 1$ is more complicated, as we shall see in the next section.

25. Lagrangian and wave equation.

Consider, finally, the case $E_o = 1$. We have

$$\mathcal{F}(1,\bar{0}) = \mathcal{D}(1,\bar{0}) \to \mathcal{F}(3,\bar{\lambda}') \,, \quad \bar{\lambda}' = (2,2,0) \,. \tag{25.1}$$

The structure of $\mathcal{F}(3,\bar{\lambda}')$ is not known. According to the theorem quoted in Section 5, $\mathcal{D}(E_o,\bar{\lambda}')$ is unitary for $E_o \geq 7$ and for $E_o = 6$. Our subsequent results show that $\mathcal{F}(3,\bar{\lambda}')$ contains $\mathcal{D}(9,\bar{0})$ as a subquotient.

The integral (24.4) becomes, when $E_o = 1$,

$$\int_{X\times X} dw d\omega \, (\det w\omega^*)^2 \, [\det(1-w\omega^*)]^{-9/2} \phi(w)^* \, \phi'(\omega) \,. \tag{25.2}$$

Now

$$\Box^2 \, [\det(1-w\omega^*)]^{-5/2} = \text{const.}[\det(1-w\omega^*)]^{-9/2} \, (\det \omega^*)^2 \tag{25.3}$$

so (25.2) becomes, up to a constant factor, if $\phi' = \phi$,

$$\mathcal{L} = \int dw\, (\det w)^2\, \phi(w)^*\, \Box^2\, \phi(w)\,. \tag{25.4}$$

This is the invariant Lagrangian for the free, even oscillator. The free wave equation

$$\Box^2\, \phi(w) = 0 \tag{25.5}$$

is an eighth-order differential equation.

26. The reduced superfield.

We have been concerned, almost exclusively, with the even oscillator representation of sp(2n). To determine the properties of the other components of the superfield, we begin by investigating the ground state of the odd oscillator, carried by the spinor field ψ. The action of sp(2n) is the obvious extension of Eq. (12.1):

$$\mathrm{sp}(2n) \ni m \to \Delta_m \quad ,$$

$$\Delta_m \psi_a = i(\delta_m + \tau_m)\, \psi_a + i m_a{}^b\, \psi_b\,. \tag{26.1}$$

The ground state is determined by

$$\Delta_m \psi_a = 0 \ \text{ for } \ m = \begin{pmatrix} a & ia \\ ia & -a \end{pmatrix} \quad ,$$

$$\Delta_m \psi_a = \frac{3}{2}\psi_a \ \text{ for } \ m = \frac{i}{2}\begin{pmatrix} 0 & 1 \\ -1 & 0 \end{pmatrix}.$$

The first equation (in which a is symmetric) expresses the fact that the energy lowering operators annihilate the ground state, and the second equation fixes the lowest energy at 3/2. Explicitly

$$\left\{ i\begin{pmatrix} a & ia \\ ia & -a \end{pmatrix} - 2ia\cdot\partial_w \right\} \psi = 0 ,$$

$$\left\{ \frac{i}{2}\begin{pmatrix} 0 & 1 \\ -1 & 0 \end{pmatrix} + 1 + w\cdot\partial_w \right\} \psi = \frac{3}{2}\psi .$$

The general solution is

$$\psi = \begin{pmatrix} (w+1)\, q \\ i(w-1)\, q \end{pmatrix} ,$$

with q constant. Comparison with Eq. (11.1) shows that the ground state spinor field, evaluated at w, belongs to the Lagrangian plane x indexed by w. This property is evidently respected by the action of sp(2n), so we have the following result:

Theorem. The odd oscillator module consists of spinor fields ψ with the property that $\psi(w)$ belongs to the Lagrangian plane $x \ni X$ indexed by w.

Since x is maximally isotropic, this property may be expressed by the equation

$$\eta^{ab}\, \xi_a \psi_b(w) = 0 , \quad \forall\, \xi \in x , \tag{26.2}$$

or

$$[-iq(w-1) ,\; q(w+1)]^a\, \psi_a(w) = 0 ,$$

for any $q = (q_1,\ldots,q_n)$.

Definition. We say that a spinor field ψ is "reduced" if at every w, $\psi(w)$ belongs to the plane indexed by w.

Note that the property of being reduced is a necessary but not a sufficient condition for ψ to belong to the odd oscillator module.

We shall now show that we can restrict ourselves to reduced spinor fields. First, let us extend the last definition to multi-spinors in the obvious way; thus (A_{ab}) is reduced if the plane that it defines at w lies in the Lagrangian plane indexed by w. Now consider the variation (3.9),

$$\delta_a \psi_b = i\left(\kappa_{ab} + \frac{1}{2}\eta_{ab}\right)\phi + A_{ab},$$

effected by applying the osp(2n/1) generator Q_a to the superfield. Suppose that A is reduced, but ϕ and A otherwise arbitrary. [Thus ϕ need not belong to the oscillator module.] Then $\delta_a\psi_b$ is reduced,

$$[-iq(w-1)\,,\ q(w+1)]^b \left(\kappa_{ab} + \frac{1}{2}\eta_{ab}\right)\phi$$

vanishes identically. This is verified by direct calculation, with

$$\frac{1}{2} m^{ab}\,\kappa_{ab}\,\phi = (\delta_m + \tau_m)\,\phi\,, \quad m \in sp(2n)\,.$$

This suggests, and explicit calculation confirms, the following result.

Theorem. The space of reduced superfields (being superfields for which all the multispinor components are reduced) form an osp(2n/1) module for the action determined by (3.7) and (3.3).

References.

The physical and especially the mathematical literature on which this paper is based is too vast for complete referencing. The following short list includes only books and papers that were actually consulted during the execution of this work.

1. N. Bourbaki, "Groupes et Algebras de Lie," Hermann, Paris 1975; Chapters 4-8.
2. T. Enright, R. Howe and N. Wallach, "A Classification of Unitary Highest Weight Modules" in "Representation Theory of Reductive Groups," P. C. Trombi editor; Birkhäuser 1983.
3. M. Kashiwara and M. Vergne, Inv. Math. 44, 1 (1978).
4. H. Rossi and M. Vergne, Acta Math. 136, 1 (1976).
5. J. Leray, "Analyse Lagrangienne et Mechanique Quantique," R.C.P. 25, Series de Math. Pures et Appliqués, IRMA, Strasbourg 1978.
6. V. Guillemin and S. Sternberg, "Geometric Asymptotics," Am. Math. Soc Math. Surveys No. 14, Providence 1977.
7. R. J. Blattner, "The Metalinear Geometry of Non-Real Polarizations," in "Differential Geometrical Methods in Mathematical Physics," Bonn 1975. (Lecture Notes in Mathematics No. 570, Springer-Verlag, 1977.
8. Seminaire H. Cartan, 10 me année: 1957-58, Vol. 1.
9. L. K. Hua, "Harmonic Analysis of Functions of Several Complex Variable in the Classical Domains," Transl. Math. Mon., Am. Math. Soc., Providence 1963.
10. I. M. Gel'fand, M. I. Graev and I. I. Piatetskii-Shapiro, "Representation Theory and Automorphic Forms," W. B. Saunders Co., Philadelphia 1969.

INDEX

Adjoint action 34, 35, 70, 227
Anti de Sitter space 4
Araki's theorem 76
Auxiliary fields 5, 52, 226

Berezin integration 17, 31

Causal structure 4, 6
Campbell-Hausdorff formula 20, 22, 36
Central charge 96, 103
Chirality 10, 69, 84, 106, 110, 119
Cohomology 91, 93
Colored singletons 10, 131, 147, 150
Conformal field theory 164, 231-255
" group 105
" invariance 2, 7, 10, 11, 136
" QED 124, 129, 135, 139
" supersymmetry 7, 8, 151-159
Compactified Minkowski space 139, 218, 232
Complex structure 211
Cosmological constant 3, 7, 68, 136
Covariant derivative 10, 80, 145

Degree of representation 134
de Sitter chirality 10, 104, 106, 110, 119
" group 3
" group representations 126
" QED 127, 129, 136-138
" space 3, 4, 6, 7, 68, 69, 86, 113, 117, 132, 136
" superfields 69-110
" supergravity 76, 150
" super QED 140-144
" supersymmetry 91-94, 103, 104, 108-110, 129, 140
Di 3
Dirac cone 139, 198, 231-235, 248
" operator 10, 51, 80, 81-105
" supermultiplet 12, 148, 149

Electroweak theory 1, 68
Extended super QED 124, 144-150
" superconformal QED 124, 154-159
" supergravity 124, 130

Field quantization 5, 10
Fierz-Pauli program 7, 88, 89
Finite reduction 17, 28
Flat space limit 4, 84, 87, 89, 94, 103, 129, 138
Forward tube domain 166, 209, 222, 239-243
Functions on supergroups 9, 17, 18, 45

Gauge fixing 138, 139
" modes 91, 110, 127, 140
" theories 4, 7, 108, 110, 127
Grassmann algebra 8, 17, 20, 27, 31, 43, 70, 111
" space 95, 112, 114
Group theory, resistance to 68
" " , why? 5
Gupta-Bleuler quantization 5, 9, 128
" triplets 76, 77, 128, 129, 138, 142, 145, 147, 152

Haar measure 9, 32, 33, 41, 44
Heisenberg algebra 55, 170
Helicity 4, 7, 87
Hermitian form 9, 17, 18, 30, 32, 33, 46-52, 76
Higgs-Kibble mechanism 10, 124
Highest weight 6, 18, 52, 61, 71-77, 83, 89, 93, 98-108, 142, 144, 146, 172, 182, 191

Indecomposable, see nondecomposable
Indefinite metric 5, 9
Induced gravity 124
Induced representations 7, 9, 17, 28, 29, 30, 45-52, 70-104, 141, 178
Infrared 1, 4, 131
" regularization 4, 126, 132, 135, 136
Integrable representations 16, 25
Invariant measure 9, 17, 18, 33, 43, 45, 46, 60
" Lagrangians 6, 84, 94, 96, 105, 231, 261
" tensors 110, 117-120
" quantization 5, 52, 93, 129
" two-point functions 97
Involutive antiautomorphism 48
Irreducible representations 1, 6, 9, 17, 28, 30, 71, 85, 126
Isotropic form 197

K-representation 30
K-structure 185
Kaluza-Klein theories 2, 10, 11, 19, 68, 166, 210

Lagrangian subspaces 197
Large ideals 105, 106, 173, 209, 238
Left translations 32, 35, 39-45, 48, 57
Lie superalgebra 19, 16-33, 55
Local group 20
Lorentz condition 92, 105, 127

Masslessness 2, 7, 9, 12, 52, 68, 73, 149, 164, 198
Minimal coupling 109
Minimal weight, see highest weight

Neutrinos 2, 152
Nondecomposable representations 4, 5, 7, 29, 51, 52, 73, 75, 93, 105, 127, 130, 132, 159, 188, 251

Oscillator family 184
" representation 55, 147, 148, 164, 170, 171, 177, 178, 196-230, 256
osp(4/1) 7, 33, 70-94, 102, 104-110, 112, 140
osp(4/2) 84, 94-106, 114, 144-149
osp(4/6) 150
osp(4/N) 131, 144-150
osp(8/1) 11
osp(2n/1) 9, 18, 37, 54, 55, 164-179
osp(2n/N) 111, 117

Parameterization 21, 33, 199
Particle interpretation 3
Phase space 53, 167
Phonons 53, 55
Poincaré-Birkhoff-Witt theorem 19
Poisson bracket 42, 44, 167
Positive energy 3
" " representations 3, 61, 71, 98, 126, 134, 139, 151
Preons 124
Projective super cone 114

Quantization 5, 52, 55, 93, 129, 164
Quasi-invariant measure 233
" -unitary representation 50
Quarks 12

Rac 3
Reduction points 64, 73, 179, 187, 208, 257
Relaxed hypermultiplet 106
Representations of superalgebras 25, 56
" of supergroups 25-33
Right translations 37, 38-45

Scalar modes 93
Singletons 3, 11, 69, 73, 126, 131, 134, 165
Solitons 124
so(3,2) 3, 69, 137, 139
so(4,2) 139
Space time 1, 210, 215, 217, 222, 230
Spin representation 148
" waves 124
Spinor superfield 106-110
Sp(4/R) 69, 72-94, 106-108
Sp(2n/R) 54, 61, 167, 179
" homogeneous space 195
Structural identities 35, 36, 39, 59
su(2,2) 8, 11
su(2,2/1) 8, 33
su(2,2/N) 154
Subsidiary condition 88, 230
Super cone 112, 117
" conformal QED 124, 129, 131, 151-154
" electroweak 124, 131
" de Sitter Maxwell equation 91
" de Sitter symmetry 3
" determinant 39, 43
" field 6, 9, 10, 31, 33, 60, 79-120, 140, 153, 174
" " representation 31, 52
" gravity 2, 10, 124, 143
" group 8, 9, 11, 16, 17, 20, 24, 25-33
" " , adjoint 24
Super group, connected 24
" " , covering 24
" " , formal 19
" " , local 8, 16, 19, 21, 23, 33
" " , real 48
" " , representation 25, 26, 27
" QED 10, 11, 140-144
" space 6, 95, 112, 114
" strings 2

Super symmetry breaking 2
" " , manifestations 2, 11
" " , prospects 11
" " , successes 1, 2
" " , unification 1
" " , why? 1
" Weyl equivalence 193
" Yang-Mills theories 2, 146-151
Symmetric algebra 17, 27, 45, 70, 142
" domains 166, 209
Symmetry breaking 126, 127, 131, 136, 222, 240
Symplectic form 5
" structure 52, 129

Twistor program 2, 164

Ultraviolet 1
Unification 1
Unitarizable representations 8, 16, 18, 26, 30, 51, 55, 61-64, 71, 85, 98, 179, 192
Universal enveloping algebra 19, 20, 21, 32

Vacuum ghost 10, 93, 124, 127, 129-131, 135, 136

Wave equations 84-91, 101, 104, 105, 117, 139, 142
Weight diagrams 74, 77, 78, 90, 100, 107, 258
Weinberg-Salam theory 1, 2
Wess-Zumino multiplet 89
Weyl quantization map 170
Witten index 134

X 197

Zero-center modules 127, 130, 132, 135, 138-159
Zuckerman translation functor 130